Spaces of Vernacular Creat

Creativity has become part of the language of regeneration experts, urban planners and government policy makers attempting to revive the economic and cultural life of cities in the twenty-first century. Concepts such as the creative class, the creative industries and bohemian cultural clusters have come to dominate thinking about how creativity can contribute to urban renewal. *Spaces of Vernacular Creativity* offers a critical perspective on the instrumental use of arts and creative practices for the purposes of urban regeneration or civic boosterism.

Several important contributions are brought into one volume to examine the geography of locally embedded forms of arts and creative practice. There has been an explosion of interest in both academic and policy circles in the notion of creativity, and its role in economic development and urban regeneration. This book argues for a rethinking of what constitutes creativity, foregrounding non-economic values and practices, and the often marginal and everyday spaces in which creativity takes shape. Drawing on a range of geographic contexts including the U.S., Europe, Canada and Australia, the book explores a diverse array of creative practices ranging from art, music and design to community gardening and anti-capitalist resistance. The book examines working-class, ethnic and non-elite forms of creativity, and a variety of creative spaces, including rural areas, suburbs and abandoned areas of the city. The authors argue for a broader and more inclusive conception of what constitutes creative practice, advocating of an approach that foregrounds economies of generosity, conviviality and activism. The book also explores the complexities and nuances that connect the local and the global and finally, the book provides a space for valuing alternative, marginal and displaced knowledges.

Spaces of Vernacular Creativity provides an important contribution to the debates on the creative class and on the role and value of creative knowledge and skills. The book aims to contribute to contemporary academic debates regarding the development of post-industrial economies and the cognitive cultural economy. It will appeal to a wide range of disciplines, including geography, applied art, planning, cultural studies, sociology and urban studies, plus specialised programmes on creativity and cultural industries at undergraduate and postgraduate levels.

Tim Edensor teaches Cultural Geography at MMU. He is author of *Tourists at the Taj* (1998), *National Identity, Popular Culture and Everyday Life* (2002) and *Industrial Ruins: Space, Aesthetics and Materiality* (2005). He is currently researching landscapes of illumination, geographies of rhythm and urban materiality.

Deborah Leslie is an Associate Professor of Geography at the University of Toronto. She is interested in the role of cultural industries in urban economic development, and has done research on a range of industries including design, fashion, art, furniture, advertising and more recently the circus. She has related research interests in the creative city initiatives and urban governance, and in the geography of commodity chains. She is author of a number of publications relating to these topics.

Steve Millington is a Senior Lecturer in Human Geography at Manchester Metropolitan University. His research interests include landscapes of illumination, vernacular creativity and geographies of play. He is co-author of *Cosmopolitan Urbanism* (2006) and has recently published journal articles in *Global Networks and Sociology*.

Norma M. Rantisi is an Associate Professor in the Department of Geography, Planning and Environment at Concordia University (Canada). She is author and co-author of numerous articles on the themes of fashion design, the cultural economy and policies governing design in urban settings. She has co-edited two special journal issues: one for *Environment and Planning A* on the creative economy and one for *The Journal of Economic Geography* on relational economic geography.

Routledge Studies in Human Geography

This series provides a forum for innovative, vibrant, and critical debate within Human Geography. Titles will reflect the wealth of research which is taking place in this diverse and ever-expanding field. Contributions will be drawn from the main sub-disciplines and from innovative areas of work which have no particular sub-disciplinary allegiances.

Published

Not yet published:

Spaces of Vernacular Creativity

Rethinking the cultural economy

**Edited by
Tim Edensor,
Deborah Leslie,
Steve Millington and
Norma M. Rantisi**

Routledge
Taylor & Francis Group

LONDON AND NEW YORK

First published 2010
by Routledge
2 Park Square, Milton Park, Abingdon, Oxfordshire OX14 4RN

Simultaneously published in the USA and Canada
by Routledge
711 Third Avenue, New York, NY 10017

First issued in paperback 2015

Routledge is an imprint of the Taylor & Francis Group, an informa business

© 2010 Selection and Editorial matter, Tim Edensor, Deborah
Leslie, Steve Millington and Norma M. Rantisi; individual
chapters, the contributors

Typeset in Times New Roman by
Bookcraft Ltd, Stroud, Gloucestershire

British Library Cataloguing in Publication Data
A catalogue record for this book is available from the British Library

Library of Congress Cataloguing in Publication Data
Spaces of vernacular creativity: rethinking the cultural economy /
eds. Tim Edensor ... [et al].
    p. cm. Includes bibliographical references and index.
    1. Intellectual capital. 2. Creative ability—Economic aspects.
    3. Professional employees. I. Edensor, Tim, 1957–
    HD53.S725 2009
    331.25–dc22   2009017534

ISBN 13: 978-1-138-98271-0 (pbk)
ISBN 13: 978-0-415-48095-6 (hbk)

Contents

Illustrations

Figures

Tables

Maps

Contributors

Alison Bain is an Associate Professor of Geography at York University (Canada) who studies contemporary Canadian urban and suburban culture. Her work examines the relationships between artists, cities and suburbs, with attention to questions of identity formation and urban change. Her current research focuses on cultural production and creative practice on the margins of Canada's largest metropolitan areas.

Chris Brennan-Horley is a postgraduate student at the University of Wollongong (Australia). His research interests include the geography of cultural industries and the application of GIS technologies to qualitative, cultural research. His PhD explores the use of mental maps and GIS for mapping the creative economy (with recent publications in *Environment and Planning A* and *Media International Australia*).

Ava Bromberg is a PhD candidate in UCLA's Urban Planning Department (USA). Her current research and practice concern innovations in resident ownership and governance of commercial real estate, public goods, and self-organised cultural spaces. Ava is co-organiser of the Just Space(s) exhibition and symposium series, co-editor of *Belltown Paradise/Making Their Own Plans*, and former managing editor of *Critical Planning*.

Jean Burgess is a Research Fellow at the ARC Centre of Excellence for Creative Industries and Innovation, Queensland University of Technology (Australia). She works in cultural studies and internet research, focusing particularly on user-created content. She is the co-author of *YouTube: Online Video and Participatory Culture* (2009), and has developed partnerships with cultural and community organisations around the uses of digital storytelling for cultural participation.

David Crouch is Professor of Cultural Geography in the Culture, Media and Communications Unit at the University of Derby (UK). He is the author/editor of many research papers, books and book chapters on visual culture, media, leisure/tourism geographies and cultural geography, and is an exhibiting artist. He is currently writing two new books *Flirting with Space: Thinking Landscape Relationally, Cultural Geographies* and *Flirting with Space: Journeys and Creativity*.

Tim Edensor teaches Cultural Geography at Manchester Metropolitan University (UK). He is author of *Tourists at the Taj* (1998), *National Identity, Popular Culture and Everyday Life* (2002) and *Industrial Ruins: Space, Aesthetics and Materiality* (2005). He has also written widely on tourism, urban and rural geographies, mobilities, and working-class identities. He is currently researching landscapes of illumination, geographies of rhythm and urban materiality.

Graeme Evans is Professor of Urban Cultures and director of the Cities Institute at London Metropolitan University (UK). He has undertaken national and international research projects into creative spaces, cultural planning and culture and regeneration for cultural ministries, agencies and the OECD. His publications include *Cultural Planning: An Urban Renaissance?*, *Designing Sustainable Cities* and over 75 journal articles and book chapters.

Chris Gibson is Associate Professor in Human Geography at the University of Wollongong (Australia). His research interests focus on creative industries (especially music), regional development and social justice. Recent major projects have examined festivals in rural Australia; creative industries in postcolonial cities; and cultural planning with disadvantaged communities. He is the co-author (with John Connell) of *Sound Tracks: Popular Music, Identity and Place* and *Music Festivals and Regional Development*.

Brian J. Hracs is a PhD candidate in the Department of Geography and Planning at the University of Toronto (Canada). His dissertation examines the nature of labour in the creative economy, with a particular focus on musicians in Toronto. His article 'Building Ontario's Music Economies' was recently included in the Ontario Government report 'Ontario in the Creative Age'.

Deborah Leslie is Associate Professor and Canada Research Chair in the Cultural Economy at the University of Toronto (Canada). She is author and co-author of numerous articles on creativity, urban-economic development and cultural policy, and commodity chains. Most recently she has been conducting research on the social dynamics of innovation in fashion, art and design. She recently co-edited a special issue of *Environment and Planning A* on geographies of creativity.

Heather E. McLean is a PhD student in the Faculty of Environmental Studies and an Executive Member of the City Institute at York University (Canada). Her current research explores the intersections of community-based art, relational aesthetics and performance with gentrification and place-marketing dynamics. She is also interested in the potential of performance for engaging residents in critical dialogue about everyone's right to the city.

Ann Markusen is Professor and Director of the Project on Regional and Industrial Economics, Humphrey Institute of Public Affairs, University of Minnesota (USA). She is an expert in urban/regional development, currently focusing on arts and culture, and her major recent works include *Reining in the Competition for Capital*, *Crossover: How Artists Build Careers across Sectors* and *Artists' Centers: Impact on Careers, Neighborhoods and Economies*.

Paul Milbourne is Professor of Human Geography and Planning at Cardiff University (UK). His principal research interests relate to the geographies of poverty, welfare and the environment. He is currently researching community gardening in urban neighbourhoods. Recent books include *International Perspectives on Rural Welfare, International Perspectives on Rural Homelessness* (with Paul Cloke) and *Rural Poverty*.

Malcolm Miles is Professor of Cultural Theory at the University of Plymouth (UK). He is author of *Urban Utopias* (2008), *Cities and Cultures* (2007) and *Urban Avant-Gardes* (2004). His current research centres on a reconsideration of critical theory, especially the work of Herbert Marcuse, for understandings of contemporary culture.

Steve Millington is a Senior Lecturer in Human Geography at Manchester Metropolitan University (UK). He has co-authored two special issues on everyday life and mobilities for *Environmental Planning A* and *Social and Cultural Geography*, and co-edited the book *Cosmopolitan Urbanism*. Recent publications concern branding and football (*Global Networks*) and domestic Christmas lights (*Sociology*) (both with Tim Edensor). His current research explores creative uses of outdoor illumination.

Tracey J. Potts is a lecturer in Critical Theory and Cultural Studies at the University of Nottingham (UK). Her research interests include material culture, lifestyle and everyday life. She has recently published articles on clutter, floral tributes and furniture design, and is in the process of completing a book on kitsch (with Ruth Holliday).

Norma M. Rantisi is an Associate Professor in the Department of Geography, Planning and Environment at Concordia University (Canada). She is author and co-author of numerous articles on the themes of fashion design, the cultural economy and policies governing design in urban settings. She has co-edited two special journal issues: one for *Environment and Planning A* on the creative economy and one for *The Journal of Economic Geography* on relational economic geography.

Bas van Heur is a postdoctoral researcher at the Faculty of Arts and Social Sciences, Maastricht University (The Netherlands). He is a geographer and cultural historian by training and his research contributes to debates on the creative industries, cultural heritage and urban development. His current research aims to broaden the debate on innovative cities by investigating the cultural and technological constitution of innovation.

Jim Walmsley is Professor of Geography and Planning at the University of New England, Armidale (Australia). His research interests include tourism, leisure and lifestyle and their impact on regional culture, demography and economy. He is currently chief investigator on collaborative research projects which investigate cultural planning, community festivals and youth out-migration in rural Australian regions.

Acknowledgements

The origin of this edited collection lies in a series of conference and panel sessions entitled 'Spaces of Vernacular Creativity' held at the Association of American Geographers Annual Conference in San Francisco, 2007. The editors would like to thank all the presenters, panellists (David Bell, Susan Christopherson, Chris Gibson, Tom Hutton, Ann Markusen) and all those in the audience for contributing to a rich and rewarding day of discussion.

We would also like to thank all the contributors of chapters to this book for their efforts in meeting deadlines and responding to editorial requests.

We would like especially to thank the Manchester Institute of Social Spatial Transformation (Manchester Metropolitan University) for all the support we have received, without which this book would not have been written.

Finally, we would like to extend our thanks to the following people, who in various ways, have made this collection possible: Jon Binnie, Bethan Evans, Julian Holloway and Mark Jayne. A special cheer for our immediate families for once again putting up with it – especially Sam, Josh, Arthur, Andrew, Emily and Rimona.

1 Introduction

Rethinking creativity: critiquing the creative class thesis

Tim Edensor, Deborah Leslie,
Steve Millington and Norma M. Rantisi

Creativity is now a central concept for regeneration experts, urban planners and government policy makers who are attempting to revive the economic and cultural life of cities in the twenty-first century. For local policy makers, a key to the economic recovery rests upon the successful development of creativity and a creative class (Florida, 2002). These ideas have been accepted, almost uncritically, by city authorities around the globe, who are intent on promoting creative quarters, clusters and networks, 'cool' cosmopolitan neighbourhoods, and the 'necessary' pre-conditions for the arrival of this creative class.

In this book, we argue that discourses of the creative city privilege particular notions of creativity, producing a hierarchal ordering which champions specific forms of urban development. In this calculus, places are ranked against one another, creating attractive 'hot spots' – invariably downtown areas and cultural quarters – but also, implicitly, their spatial 'other': cultural deserts devoid of coolness. It is timely, therefore, to challenge such readings of creativity. We offer a critical perspective on the instrumental use of creativity for urban and non-urban regeneration and economic development. In particular, we examine how notions of a creative class construct restrictions around who, what and where is considered 'creative' and argue that an understanding of vernacular and everyday landscapes of creativity honours the non-economic values and outcomes produced by alternative, marginal and quotidian creative practices, and has the potential to move us toward more holistic, diverse and socially inclusive creative city strategies. First of all, we situate current conceptions of creativity within the recent neoliberal hegemony. Second, we discuss how creativity might be conceived otherwise and third, we discuss a broader spatial terrain within which creativity unfolds.

The rise of a creative city agenda in a neoliberal age

Much of the recent interest in creativity is linked to the fundamental economic restructuring since the early 1990s and the promise that creativity holds for bolstering economic development. Since the 1970s, the advent of globalisation – along with advances in telecommunications and transportation systems and a loosening of trade barriers – has levelled the economic playing field and altered the basis of competitive advantage. As mass markets become increasingly saturated, firms are seeking to differentiate their products and to compete on the basis of signs and symbols, as opposed to the physical attributes or functionality of commodities (Lash and Urry, 1994). Scott, for example, contends that a critical arena in which competition is being waged today is in the development of marketable outputs whose qualities 'depend on the fact that they function at least in part as personal ornaments, modes of social display, aestheticised objects, forms of entertainment and distraction, or sources of information and self-awareness' (1997: 323), namely as artefacts whose psychic gratification to the consumer is high, relative to their utilitarian purpose. From this standpoint, creativity is defined in terms of aesthetic experimentation or 'innovation', whereby planned obsolescence rests on a marrying of the cultural with the technical and commercial (Jameson, 1984). If novelty is the economic order of the day, creativity is perceived as the key ingredient.

The centrality of creativity – and by extension, arts and culture – to economic competitiveness is not only extended to marketable outputs; it has also become enmeshed in recent efforts at place making. Here, the emphasis is on the construction of spectacular spaces of culture and consumption, such as festival marketplaces, creative quarters and cultural facilities designed by world-renowned, 'star' architects (Evans, 2003; Bell and Jayne, 2003; Hannigan, 1999; McNeill, 2008). These forms of place making are seen to enhance the cultural capital attached to the commodities emanating from a city. Molotch, for example, suggests that 'the positive connection of product image to place yields a kind of "monopoly rent" that adheres to places, their insignia, and the brand names that may attach to them' (1996: 229). Scott (1999) similarly suggests that the temporal and spatial qualities associated with particular places are grafted onto the products produced in them and that these goods come to define their places of origin. Under this logic, creative industries foster the development of identities which feed back into new rounds of production.

This fusion of culture and creativity with the economy has been coupled with changes in state policy and support for the arts. Lily Kong (2000), for example, charts a transition from an era where culture and the arts were supported according to social and cultural rationales – under a banner of community development or 'art for art's sake' – to an age in which urban policy makers have foregrounded the role of creative industries in economic development and urban renewal (see Evans, this volume). The 1980s marked a significant turning point as governments began adopting a neoliberal approach that involved an extension of market principles to state policies as a means to contend with global competition, industrial decline and market volatility (Harvey, 1989). With respect to the arts, public subsidies were increasingly viewed as a form of welfare and fell victim to broader cuts in public spending as governments were becoming less managerial and more entrepreneurial in their approach to service provision. This was particularly the case for cities, which faced less support from higher levels of government and were expected to take on more responsibility for leveraging their own revenues (Brenner and Theodore, 2002). In light of fiscal constraints and inter-urban competition for capital, creative industries were increasingly valued in terms of their ability to foster a new image for the city that would enhance its economic competitiveness and attract talent, shoppers and tourists. Under this new economic agenda, creative industries were viewed as a vehicle through which cities could distinguish themselves, as well as the goods and services they produced (Peck, 2005).

The broader political-economic context in which creativity has assumed economic significance cannot be divorced from prevailing academic discourses that have sought to capture and highlight the economic dimensions. The two most influential sets of works that have shaped creative industry policy, to date, include the publications of Landry and Bianchini (1995) and Landry (2000) on 'creative cities' and the publication of Richard Florida (2002) on the creative class. The 'creative city' concept was first introduced by Landry and Bianchini (1995) in their book entitled *The Creative City* and then refined and repackaged for a policy audience in *The Creative City: A Toolkit for Urban Innovators* by Landry (2000). The basic tenet of this concept is that cities are facing immense challenges with the transition from an industrial to a post-industrial era and need to be creative in thinking of solutions to urban problems. Urban development and renewal, they contend, must move beyond investments in the physical attributes of place and infrastructure – and consider the role that local cultural resources can play in urban revitalisation. Such cultural resources can shape the prospects for enhancing local connections as well as projecting a distinctive image for cities at a time when, it is contended, culture is becoming increasingly homogenised around the globe.

The creative class concept, developed by Richard Florida (2002), also acknowledges the shift from an industrial to a post-industrial society and from a society based on the production of goods to one based on the production of ideas. Florida charts the rise of a new class of workers who are said to embody the knowledge demanded by this new economic order, a class that encompasses those working in science, research, law, education and training, arts, culture and technology. Florida presents the creative class as 'the dominant class in society' (2002: ix) in terms of its influence in developing new ideas, products and lifestyles. These workers are seen to be highly mobile and attracted to a diverse, tolerant environment open to unconventional thinking and new ideas. Florida's battle cry, 'Be Creative or Die', has become a highly influential maxim for industry leaders and politicians (Dreher, 2002) and a key aim has been to produce more dynamic, entrepreneurial and cosmopolitan places that also embody liberal values towards difference, as grounded in Florida's mantra of the three T's: technology, talent and tolerance. Florida's arguments echo the earlier observations of Jacobs (1961) about the role of diversity in the economic prosperity of cities, and in terms of policy orientation there are strong parallels with the creative city concept in which the development and promotion of local cultural resources are championed as a way of enticing and retaining this class of people.

Limitations of the creative city/class agenda

The influence of the creative city and creative class concepts has spread far and wide, and this popularity suggests that this is an idea whose time has come. However, we argue that such policy responses represent an instrumental capturing of creativity around a particular neoliberal economic and political ideology, related to fostering labour market participation, civic boosterism and competitiveness. In practice, this has translated into a privileging of particular entrepreneurial practices, urban locales and a meritocratic class.

Artist as entrepreneur

First, the creative practices associated with cultural and economic regeneration have become explicitly aligned with market-ready activities which redefine the artist as a creative entrepreneur. The concept of the 'creative city' has promoted the valorisation of particular forms of creativity, including a proclivity to promote only those cultural activities whose products are easily commodifiable in terms of intellectual property rights and copyright material, such as music and film. As a consequence, other forms of creativity that fall outside the creative city agenda tend to be marginalised. And a normative, constrained definition of the cultural economy can lead to policy decisions that further disenfranchise other types of cultural production. The privileging of an entrepreneurial notion of creativity is captured by Montgomery, who states that 'creativity should be certainly be [*sic*] encouraged, but not without a concomitant commitment to enterprise, risk-taking and wealth creation' (2005: 342–3). Such a stance precludes the consideration of 'alternative creativities' whose cultural products are not so easily commodifiable (Gibson and Klocker, 2005). In his critique of the creative class script, Jamie Peck (2005) demonstrates how the instrumental deployment of creativity in urban policy is imbued with the moral imperatives of the neo-Right, whereby the fostering of individual creativity is tied to the fulfilment of neoliberal economic priorities. The *Cox Review of Creativity in Business in the UK* (2005), for example, firmly equates creativity with enterprise and innovation: '"creativity" is the generation of new ideas – either new ways of looking at existing problems, or of seeing new opportunities, perhaps by exploiting emerging technologies or changes in markets' (Cox, 2005: 2). Redefining the artist as 'entrepreneur' and risk taker has serious consequences for work conditions in the new creative economy. Whether at the scale of the individual, a single business, or even a whole city, the necessity of being prepared to act and respond creatively for your own survival is emphasised.

While popular accounts of independent creative workers often project a romanticised picture of self-fulfilment and independence from routine nine-to-five work schedules, academic scholars have drawn attention to how such work is characterised by high levels of insecurity (McRobbie, 1998; 2005; Gill, 2002). As McRobbie's research on independent designers illustrates, creative workers often move from one project to the next, work in isolation and need to make a name for themselves by building up their own brand. Moreover, since much of the work they take on is freelance work for large corporations, as casualised labour, they must assume the risks and costs of their creative output. As a consequence, such workers become self-disciplining, neoliberal subjects, assuming full responsibility for any failings, working long hours to attain success and looking after their own self-interest (McRobbie, 2005). Banks (2009) further points out how distinctions between work and leisure become blurred for such people by virtue of a time squeeze generated by a concern with finding productive creativity in pastimes and pleasures.

The creative city script thus presents an idealised image of an entrepreneurial creative subject, neglecting the power relations, discipline and risks that confront members of the so-called creative class. Such a conception of creativity also neglects less commodified, alternative and often more subversive forms of creativity in the city, as we discuss in more detail below.

The focus on the urban

One of the most glaring inadequacies of the creative class thesis and its many advocates is its geographical specificity. We are particularly concerned over the consensus in cultural policy and academic discourse that privileges large metropolitan centres as sites of cultural production. There is no doubt that cities have been great centres of technological and cultural innovation and that within cities risks are taken, problems raised, experiments tested and ideas generated. Historically, people gravitate to cities for employment, stimulus or the comfort of strangers and today, certain renowned cities with economic, cultural and political power continue to act as magnets for social and commercial experimentation: 'the innovative places the last time around look like being the creative places the next time around' (Hall, 2000: 648). Accordingly, certain cities such as Manchester, Barcelona and San Francisco are held up as models of cultural dynamism worthy of imitation, and undoubtedly cities such as London or New York are dynamic centres of cultural production, and possess particular advantages in terms of supporting the creative economy. However, the instrumental use of arts and culture in strategies for urban renewal continues to venerate the metropolis as an especially unique environment, conducive to creative production and consumption to the exclusion of other spaces. Such policies fail to appreciate the complexities of urban living in a networked age. The developments of exurbia, networked culture and enhanced mobilities suggest that the experience of urban living is changing rapidly. And as Miles and Paddison (2005) argue, the tendency of contemporary approaches of cultural policies to overemphasise the centrality of large metropolitan areas as the principal realms for creative production and consumption implicitly denigrates the significance of peripheral, marginal and non-urban spaces. Florida (2002), for example, has encouraged a fascination with the ranking of cool cities, where the creative industries are said to flourish, into league tables. However, for every top ten cool city, there is also a 'crap town' (Jordison and Kiernan, 2006), and spaces beyond the urban or non-metropolitan in which other forms of cultural production are unknown, ignored and trivialised.

Cultural regeneration strategies not only privilege metropolitan areas, but within these areas tend to fetishise particular aestheticised spaces of production and consumption, such as the gentrified urban centre or bohemian cultural enclave. These spaces are often characterised by state-led spectacular flagship projects and events, together with the development of distinctive cultural quarters. Strategies to cultivate such spaces involve urban design, an improved public realm and image campaigning to imprint a new identity on the urban landscape that is appealing to both local artists and the consumers of culture – and perhaps more importantly, institutional investors or tourists (Bell and Jayne, 2004; Gibson and Kong, 2005). Moreover, in line with much of the recent literature and the neoliberal policy focus on the relation between clusters and innovation (see Porter, 1998 for example), these centres generally represent a concentration of creative industries, as the co-location of activities is seen to enhance knowledge exchange and the generation of new ideas and practices, not to mention the potential reputation effects for a distinctive cultural quarter.

To date, however, there is a paucity of empirical evidence demonstrating that clusters provide sustainable vehicles for urban regeneration (Evans, 2005), and according to Zukin (2006), Román (2006) and Vicario and Martínez-Monje (2006), the concentration or agglomeration of creative industries may actually do more harm than good. First, a city's comparative advantage may be undermined through the standardisation of cultural policy models, the serial reproduction of cluster forms and subsequent homogeneity of the built environment in terms of urban design, creative content and public realm. Indeed, given the tendency of local policy makers to reproduce anything that looks remotely successful, the twenty-first century has witnessed a global spread of similar cultural policy regeneration initiatives. Second, the concentration of resources and investment within clusters serves the exclusionary power of gentrification, creating attractive and desirable neighbourhoods which may force out the original urban pioneers through inflated capital and rental values. Zukin (1989; 1995) has shown, with her seminal work on Soho (New York), how culture-led renewal vanquished the conditions that led to the formation of artistic milieus in the first place. Third, the sheer scale of investment in major artistic flagship projects can divert resources away from much-needed investment in social development, as well as reproduce spatial inequalities in the geography of artistic and creative production.

As we discuss more critically below, we argue for a broader conception of creativity, one that recognises its practice outside the city, the downtown and the cultural quarter. The prioritisation of such spaces also has a class-based dimension, which we now highlight in the final section.

The priorities and politics of the middle class

In current conceptualisations, members of the creative class possess the appropriate qualities for producing creative work or products of value. In terms of their spatial preferences, they are drawn to locales with art galleries, chic shopping areas, heritage, museums, cafés and a reputation for cosmopolitanism and liberal attitudes towards sexuality, gender and other forms of difference. Yet, such an emphasis is shrouded in a particular set of middle-class values, and the implication persists that differently positioned social groups lack the necessary creative skills, cultural tastes and competencies to effectively operate within the creative economy, and even more, that there is a creative class – and therefore other classes that are not creative. The most uncreative class here seems to be the abject working class, but, as we discuss below, the suburban middle class also falls into this category.

The problems with this emphasis are twofold. First, as Haylett (2003) suggests, a state of being uncreative is redefined as a problem for the state to deal with. Creativity in this context, therefore, becomes a discursive weapon to further problematise non-middle-class values and peoples. Second, and relatedly, there is a failure on the part of contemporary cultural policy to engage working-class communities within visions of cultural regeneration. Cities lacking the cultural capital bestowed by the presence of art galleries and upscale cafés have a more difficult job in attracting that class of creatives who aspire to be surrounded by such facilities and philosophies. Cultural regeneration, it would seem, grinds to a halt when it encounters the type of place Evans and Lowery (2006) have described as 'roll up your sleeves kinda towns', gritty working-class cities, such as Stoke-on-Trent or Hull, Gary or Cleveland, or Hamilton. If you cannot 'Landrify' Cleveland or attract the creative classes to the Potteries, then what sort of cultural policy would work in those locales?

As presently defined, then, the creative class produces an explicit, reconfigured version of the old hierarchy between 'high' and 'low' cultures and serves to reproduce class distinctions. Lawler (2005) demonstrates how the making of such distinctions is bound up with the contemporary formation of middle-class identities through the articulation of a distinction between those people and places said to possess cultural competence or spatial capital, often expressed as 'cool' and 'uncool' (Hannigan, 1999; also see Potts, this volume). In this context, it is crucial to understand the selective production of creative products and practices through the endeavours of a range of cultural 'experts' who, by self-definition, are trying to form themselves into a particular 'class' identified by particular forms of distinction (Bourdieu, 1976). Mike Featherstone (1991) highlights the expansion of cultural commodification through the strategies of previously marginalised sections of the middle classes to claim and mobilise forms of expertise around previously lowly, marginalised and unfashionable cultural forms and practices. The rise of the creative class reveals an intensification of the process through which areas of creativity might be privileged along with a group of 'creative' people – cultural promoters, regeneration professionals and policy makers – shaped by a habitus and particular dispositions towards arenas of creativity.

To counter these exclusionary processes, Gibson and Klocker maintain that

> Contemporary urban-social policy needs the kind of imagination that can understand something of the texture of poverty and working-class lives as ordinary and extraordinary ways of being. Without such thinking, working-class people and places can only ever be 'less than' those in whose image they are reconstructed.
>
> (2005: 101)

In this book, we examine the politics of the creative city script, and the forms of creativity that are excluded from hegemonic accounts. In the next section, we argue for a broader conception of creativity, one that recognises that it is widely distributed among a diverse set of spaces and peoples.

Rethinking creativity

Creativity is commonly conceived as the property of particular individual artists, thinkers and philosophers, exceptional, individual geniuses who are able to transcend the bounds of what is presumed to be ordinary thought and practice. However, having outlined the deficiencies of the restricted conceptions of creativity so keenly taken up by urban regeneration planners and cultural policy makers, we suggest that creativity is far more widely distributed across hugely diverse spaces and cultural contexts. Associated with the idea of the uniquely creative individual (see Bourdieu, 1993; Becker, 1982) is the notion that creativity is characterised by novelty (for instance, in the recent work of John Tusa, see Bain, this volume). As we have discussed above, this neatly conforms to the desire of capital to foreground forms of creativity that promise the new, and so rapidly supersede the instantly unfashionable in the endless quest for new commodities and markets.

However, according to Hallam and Ingold, creativity should rather be conceived as an improvisation quality that, across all forms of cultural activity, requires people to adapt to particular circumstances. Even in apparently repetitious practices, regeneration actually takes place under circumstances that are always different. Accordingly, ideas that creativity is always innovative, producing novelty, ignore how cultural practice invariably 'entails a complex and ongoing alignment of observation of the model with action in the world' (2007: 5). This means that there 'is creativity even and especially in the maintenance of an established tradition … [for] traditions have to be worked to be sustained' (ibid.: 5–6). Importantly, this refutes the idea that tradition is the antithesis of creative innovation. Instead, as Giddens also argues (1994), tradition is persistently reproduced through active regeneration, refuting the suggestion that, through history, continuous time is punctuated by moments of innovation and inspiration. The temporality of creativity here is located in the adaptation of pre-existing ideas and intentions to a fluid present and is thus constantly evident 'in the circulations and fluxes of the materials that surround us and indeed of which we are made' (ibid.: 11).

Ideas about the creative production of novelty suggest the valorisation of rapid and dynamic change, in contradistinction to the routinised and habitual everyday within mundane realms in which change occurs at a slower pace. However, Hallam and Ingold develop their argument by maintaining that these apparently banal realms are not devoid of creativity, because 'the forward movement of keeping life going, however, can involve a good measure of creative improvisation' (2007: 7). This understanding chimes with Paul Harrison's notion of the everyday as that which is not merely characterised by dull compulsion and repetition: 'in the everyday enactment of the world there is always immanent potential for new possibilities of life' (2000: 498). Quotidian practice is open-ended, fluid and generative, concerns becoming rather than being, is a sensual experiencing and understanding that is 'constantly attaching, weaving and disconnecting; constantly mutating and creating' (ibid.: 502). The notion of the potentialities of routine also resonates with Lefebvre's insistence that the rhythms of social life never involve any exact repetition (2004).

Hallam and Ingold further dispute the aforementioned thesis that creativity is the property of a sovereign individual or gifted genius. Instead, it is relational, involving 'persons in those mutually constitutive relationships through which, as they grow older together, they continually participate in each other's coming-into-being' (2007: 6). The idea that the uniquely creative person can somehow disentangle himself or herself from the social (including the non-human) world is dismissed since, as 'it mingles with the world, the (individual) mind's creativity is inseparable from that of the total matrix of relations in which it is embedded and into which it extends' (ibid.: 9). Implicitly then, creativity is social and sociable, culturally specific and communally produced, and is located in innumerable social contexts.

In acknowledging that creativity is located in everyday, popular, vernacular culture, it is timely to revisit the work of cultural studies which in its early British incarnation radically interrogated the distinction between so-called 'high' and 'low' culture and more specifically sought to positively evaluate local, popular, subaltern and everyday cultures and subcultures, drawing, as Featherstone (1992) depicts it, a distinction between a somewhat mythical 'heroic', artistic life and the practical, situated activities within everyday culture. The blurring of these supposedly previously distinct cultural spheres is now axiomatic, through the endeavours of writers such as Dick Hebdige (1983) to demolish hierarchical cultural boundaries in pursuing Raymond Williams' anti-auratic suggestion that 'culture is ordinary' (1997). Here, we particularly remember the reassessment of creativity in the rather unfashionable work of Paul Willis, who insists that within what he calls 'common culture', 'there is a vibrant symbolic life and symbolic creativity in everyday life' (1993: 206), not a supplement supplied by conventional 'art' but 'part of the necessariness of everyday symbolic and communicative work' (ibid.: 208), that (re)produce identities, places and vital capacities of humans. More specifically, Willis draws attention to

> The extraordinary symbolic creativity of the multitude of ways in which young people use, humanize, decorate and invest with meanings their common and immediate life space and social practices – personal styles and choice of clothes; selective and active use of music, TV, magazines; decoration of bedrooms; the rituals of romance and subcultural styles; the style, banter and drama of friendship groups; music-making and dance …
>
> (ibid.: 206–7)

Willis further identities a 'grounded aesthetics' as part of the ordinary, symbolic qualities of everyday existence, embedded in sensation, fun, desire and festivity rather than synonymous with the 'cerebral, abstract or sublimated quality of beauty. While such ideas have been absorbed into mainstream social science orthodoxy and lost their radical edge, and been subject to valid criticisms that they are too optimistic, politically uncritical and culturally populist (McGuigan, 1992), they do have the great virtue of emphasising the protean and fluid nature of creativity and the agency of creative actors within all social groups. Thus suburban poetry groups, web-page makers and bloggers, knitting circles, photography and home movies, car customising, community festivals and horticultural shows can be added to Willis's list, as well as numerous other practices. A significant academic work capturing these pursuits is that by Paul Hoggett and Jeff Bishop (1986), which focuses on the collective pursuit of hobbies for enthusiasm and enjoyment rather than need. They recognise that some groups may be short lived but can provide a means by which people can share interests, engage in collective projects and develop friendships, and that such endeavours also 'consistently offer enthusiasts the opportunity to develop a sense of value and identity' through creating (ibid.: 57). Another study that identifies the values of these amateur groups is Ruth Finnegan's (1989) depiction of the plethora of amateur, grassroots music makers in the town of Milton Keynes. For these participants, argues Finnegan, musical practice was a profoundly creative process irrespective of value-laden notions of its quality or genre, and contained a large quantity of composition, whether adaptive arrangements or original across all genres. Crucially, she asserts that an 'exactness of repetition from practice into performance was not complete … Every performance, whatever the composition mode, must involve some leeway for individual creativity' (ibid.: 169) each time, and this creativity could involve individual or collective improvisation and interpretation, performative modes, lyric writing and audience participation and so on. Again, Finnegan points out how these creative pursuits are not purely instrumental ventures but are embedded in the routines of everyday life and foster sociability and conviviality, affective pleasures as well as distinction. Moreover, Finnegan identifies musical family habits and traditions as key in stimulating participation rather than class background.

Given its ubiquitous nature, vernacular creativity also possesses power to transform space and the everyday lives of ordinary people to reveal and illuminate the mundane as a site of assurance, resistance, affect and potentialities. This is something that cannot be measured or equated simply in economic terms; for example, it could provide a basis for civic identity, unity or sense of self-esteem (Burgess, 2006). Accordingly, in this volume we foreground aspects of creativity left out of this conceptualisation, highlighting the importance of vernacular and everyday forms of creativity. Vernacular creativity foregrounds the un-hip, the un-cool, and possibly the downright square, and embraces those marginal and non-glamorous creative practices excluded from arts- and culture-based regeneration. Vernacular forms of creativity are neither extraordinary nor spectacular, as Burgess notes (this volume), but are part of a range of mundane, intensely social practices grounded in a variety of everyday practices and places, as we discuss below.

In rethinking the notion of creativity, we also wish to disentangle it from economic instrumentality by arguing that there is much to be said for non-productive creativity – creative practices that merely distract the producer from the travails of every day, acts that may appear to the outsider to be nothing more than dilly-dallying, time wasting or procrastination. The value of 'unproductive' creativity has been lost in the mire of discourses that connect creativity to economic objectives and ignore practices and forms that are non-commodifiable. Our intention is to examine how non-economic values are derived through engagement with alternative and vernacular creative practices. Chapters thus examine the affectual and sensual qualities of creative activities which may appear frivolous and playful (see Crouch, this volume) and that produce community cohesion, neighbour-hood identity, self-worth, sociality, conviviality or the production of economies of generosity (Bromberg, Edensor and Millington, Milbourne, this volume). This is not to simply decry the uses of creativity for economic purposes nor to posit a crude distinction between economic and non-economic creativities, for as several chapters here show, there is often a blurring between these spheres, wherein, for example, the non-economic can produce an economic resource And we want to further rethink creativity as immanent and emergent, in a state of constant rein-vention and reinterpretation, as Hallam and Ingold (2007) emphasise. As Lefebvre (1991), de Certeau (1984) and others remind us, we should not underestimate the potential of the banal to produce alternative and resistant everyday practices that enable individuals to reclaim some autonomy or control over dislocated power.

Finally, although Allen Scott's primary motivation is to define the nature of the cultural economy and analyse the creative processes that underpin it, partic-ularly noting the potential of digital and network technologies and rises in the consumption of new cultural forms, he is also critical of the instrumental attempts to nurture economic development through the tactics we have critiqued above. He argues that such policy devices are predicated on abstract notions of crea-tivity, and instead prioritises the empirical imperative to analyse how creativity is constituted within everyday, locally embedded social situations. Accordingly, drawing on Bourdieu (1993) and Csikszentmihalyi (1997), Scott (2008) under-stands creativity as culturally situated and produced rather than the product of some transcendental individual genius situated within a particular time and space within 'creative' fields in which existing forms of knowledge, skill, habit and modes of socialisation persist. Here, we note that the acknowledgement of a more distributed sense of creativity is not only politically significant, but is also being recognised by some with a broader concern to explore the economic uses and cultural politics of a wider creativity.

Alternative spaces of creativity

As we have mentioned, one of the most glaring inadequacies of the creative class thesis is its geographical specificity, privileging downtown cultural enclaves and quarters in large metropolitan centres as sites of creativity. The champions of creative regeneration have fetishised these urban settings while ignoring forms of creative endeavour that emerge in rural, suburban, working-class, everyday and marginal spaces. The additional focus on the constitution of creative centres through the formation of actors into *clusters* introduces a further spatial restric-tion that negates more dispersed spatialities of creativity, not least, the expansion of creative networks. This book thus identifies a broader spatial contextualisation for creativity, relocating analysis away from a narrow fascination with the metro-politan centre.

A major constraint is the tendency to produce and maintain dichotomous under-standings of creativity and space through the reproduction of binary spatial distinc-tions between global/local, cool/uncool, creative/uncreative, fixed/mobile, centre/periphery and urban/rural contexts. These dichotomies abandon a more inclusive analysis of the geography of creativity, which is multiple, mutable, unpredictable and dynamic. We have already referred to the class bias in conceptions of creativity and the tendency to regard working-class estates, streets, homes, garages, sheds and gardens as devoid of creativity. Also, however, other spaces are represented as equally bereft of creative activity. Here we focus upon suburbia, alternative and marginal spaces, and everyday realms, before turning our attention to creative networks.

Suburbia and rurality

In the UK, the DETR (2000) estimates that just 5 per cent of the UK's population are rural, 9 per cent reside within urban cores, and the remaining 86 per cent reside in some suburban/urban-rural fringe context. Given that significant subcultural practices in music, fashion and publishing largely emerged from suburban settings outside of world cities, we question the extent to which creativity is the preserve of large metropolitan cities.

Suburbia is still commonly represented, in films, situation comedies and other media forms, as a site suffused with conformity, a cult of domesticity, self-enclosed individuality, mindless aspiration, indulgent consumerism, tedium and blandness. A frequent subtext is that underneath the playing out of rigid conventions lies a frustrated desire that finds an outlet in the sexual shenanigans of desperate housewives, teenage rebellion and a host of other hidden and forbidden pursuits, as famously evinced in David Lynch's film, *Blue Velvet*. A suffocating blanket of keeping up appearances and demonstrating a constrained notion of respectability, together with a presumed dearth of multicultural and cosmopolitan openness born of the desire to escape the metropolis, portrays the suburb as a cultural desert. Yet, as Roger Silverstone insists, 'suburbia is creative ... [it is] a social as well as a cultural hybrid' (1997: 6–7) and, moreover, has served as the crucible for the emergence of British punk rock as well as for the idiosyncratic domestic productions of innumerable householders in their homes and gardens. Moreover, reified notions of suburbia are dissolving as suburbia assumes ever more forms, from 'technoburb' and themed town to inner-city suburb and heritage district, and given the extension of creative processes across space. As Hracs, Burgess and Bain demonstrate in this volume, suburbia is increasingly a venue for the residence and production of those more conventionally associated with the creative arts. And in fact, disdain towards the suburban by cultural policy makers is presently under challenge within the UK, where the need for suburban regeneration and polycentric planning is now advocated by central government (ODPM, 2004).

Similarly stereotyped as lacking creativity is the rural, confirming long-held academic and popular binary understandings that it is the opposite of the city, not sophisticated, cosmopolitan, modern and liberal, but backward, traditional and offering few of the creative possibilities available in the urban. Yet, as Gibson, Brennan-Horley and Walmsley show in this volume, the rural can be the site of surprising and innovative creative projects of many kinds and Bell and Jayne (forthcoming) point to how rural in-migration from artists, down-shifters and commuters has energised the possibilities for creative production. All of the convivial pastimes and creative enthusiasms already cited are as likely to occur in rural settings, as well as a host of rural music festivals, raves and music-making activities and a variety of craft pursuits, community festivals as well as particular artistic expressions such as the production of elaborate crop circles and straw-bale art.

Alternative and everyday community spaces

Similarly omitted from the focus on the city centre as *the* realm of creativity are those interstitial spaces that permeate the city but are often seen as devoid of value. Here, we are referring to the 'wildscapes' (Jorgensen and Keenan, 2008), the urban fringe, spaces of industrial ruination and wasteland (Edensor, 2005), rubbish dumps (see Potts, this volume), the undersides of flyovers and a host of other marginal realms. Such spaces provide a site in which creative play, art practices, graffiti composition and other uses can occur without forms of urban regulation restricting what may take place. Unofficial and unheralded spaces such as these permit a playful, unselfconscious, experimental engagement with space and an opportunity to physically engage with textures, materials and other affordances away from the strictures of preferred aesthetics and value. This is particularly evident in the creation of Nek Chand's extraordinary tourist attraction adjacent to the Indian city of Chandigarh (Nek Chand Foundation). As the city was being partly demolished and rebuilt according to the designs of Le Corbusier, Chand used discarded rubble to create a motley collection of sculptures of fantastical beasts and people in a secluded wooded area on Chandigarh's outskirts, works uninfluenced by any conventional art school. Upon its discovery, the city authorities initially slated his creation for demolition but then made the enlightened decision to provide him with a workforce and a salary so that he could continue his singular vision, a decision that has produced the amazing Rock Gardens of Chandigarh. Such productions have typically been created by those outside the art mainstream and forged in mundane, domestic and marginal spaces, other examples including Simon Rodia's Watts Towers in the working-class district of Los Angeles, postman Ferdinand Cheval's Palais Idéal in his garden in the nondescript French village of Hauterives, and the Maison Picassiette, a council house covered in mosaics in Chartres, France. These astounding creations are rare, but everyday houses and gardens, communal areas and neighbourhood streets are similarly sites in which a mundane, vernacular creativity takes place.

It is impossible to account entirely for the myriad creative forms and practices that saturate the dense environment of the everyday, yet as Evans declares, 'it is the everyday lived cultural practices and experiences ... which, the evidence suggests, better represent cultural regeneration primarily through social and community-based projects' (2005: 976). We have already mentioned the less glamorous spaces of community arts that more inclusive cultural policies continue to include in regeneration strategies, the 'community embedded artistic spaces' (Markusen, 2006) (Evans, McLean, Markusen, Miles, this volume) and innovative and collective centres that offer alternative routes to creative production and pleasures (Bromberg, this volume). It is also clear that besides the ordinary community centres and parish halls within which hobbies and pastimes are enjoyed, back alleys (Milbourne, this volume), house facades (Edensor and Millington, this volume), allotments (Crouch, this volume), rubbish tips (Potts, this volume), cafés (Rantisi and Leslie, this volume) and the increasingly ordinary world of cyberspace (van Heur, Burgess, this volume) are mundane spaces in which creativity can be pursued. Rather than being dismissed as the banal opposite of bohemian cultural milieus, as abject, marginal, parochial or backward, such overlooked sites are those in which vernacular creativity imprints class and ethnic identity upon the landscape, that resonate with affective and expressive values and that articulate communal conviviality and social solidarities. The challenge for cultural policy beyond an instrumental recuperation of 'art' as an economic resource must be to develop a more reflective and inclusive position regarding the value of everyday or vernacular forms of creativity, for as Gibson and Kong insist, for many, 'participation in cultural activities is initially driven not by career development motivations, but by a personal desire to engage with the affective, emotive, cathartic dimensions of creative pursuits' (2005: 544).

Networks of creativity

Clusters are said to generate positive externalities between firms located in close proximity to one another, for example, through local exchanges of ideas, skills and technology, an observation noted by Alfred Marshall about the South Lancashire cotton trade in the nineteenth century. During the 1960s, ideas such as Perroux's Growth Pole Theory (1950) reaffirmed the importance of supporting economic development around specialised centres of technology, which influenced a range of state-led policy interventions to tackle the problems of deindustrialisation. In recent times Porter (1985) has argued that agglomerations of skills and technologies around specialised centres of production afford cities and regions unique comparative advantages over other places. These ideas are aligned with discussions about the transition to post-Fordist forms of economic development, whereby centres of expertise are seen as either advantageous nodal points within complex global networks (Sassen, 2006), or essential to the formation of multi-faceted flexible production (Piore and Sabel, 1984) or Regional Innovation Systems. It is perhaps not surprising, therefore, given that creative industries are often positioned as 'sunrise' industries, suited to the conditions of late capitalism, that clusters continue to loom large in theories of creativity.

Accordingly, a multitude of cities have tried to nurture cultural clusters or quarters supporting specialised forms of creative practice as the basis of regeneration and civic renewal. Yet, while their existence performs an important discursive function in generating positive place promotion, empirical evidence suggests that clusters are relatively weak vehicles of economic growth and development (Mommaas, 2004). Bathelt and Malmberg (2004) and Grabher (2002), among others, have written about processes that embed creative industries into localised agglomerations, but much less is understood about how creative clusters are linked to multi-scalar networks (Pratt, 2008). Our concern here, then, is how the deployment of clusters serves to establish boundaries around creativity, marking out 'creative spaces' in distinction to other, 'ordinary' spaces, and ignoring creative geographies that are socially produced activity across a range of sites and spaces in ways that are more rhizomatic or viral.

Consequently, instead of focusing upon clusters and other concepts that suggest spatial boundedness, limited porosity and a lack of openness to ideas from further afield, and in accordance with our desires to foreground the ongoing (re)distribution of creativity and develop a more open understanding of the transitory and fluid nature of creative practice, we feature networks as a more appropriate spatial context within which creative projects can be practised. For rather than conceiving of creativity as taking place among groups in hierarchical structures and bounded places, we follow Castells' assertion that networks 'constitute the new social morphology' (1996: 469), here defined as comprising a set of interconnected nodes, ties and flows (Barney, 2004) and of varying scales, geographical extent, volatility and dynamism, and duration. The network emphasises the relationality of the social, indeed insists on an ever-increasing multitude of connections and chains of relationality through which qualities such as creativity are continually redistributed. Here, Wellman's contention that communities can be depicted as 'networks of interpersonal ties that provide sociability, support, information, a sense of belonging and social identity' (2001: 227) has been extended for many by developing communications technologies and expanding mobilities. The creative currents that flow through networks thus increase the potential for new and emergent forms of activity across a range of sites and locales, as inter-scalar flows, relations and social dynamics connect local practices to wider networks of cultural and economic activity.

The developments of exurbia, networked culture and enhanced mobilities suggest that the experience of urban living is changing rapidly and is certainly becoming networked at a larger scale. Yet, as Finnegan's study shows, creativity has long been constituted in and through networks. The amateur musicians criss-crossing Milton Keynes continually recompose heterogeneous networks of greater or lesser fluidity across the city rather than reproducing a circumscribed creative community in place. Similarly, and on an increasingly large scale, the ability to sustain 'invisible networks' is crucial to the tactics of many internationally constituted new social movements, often transient, non-hierarchical and experimental in form (Melucci, 1989), and a host of affective groupings, collectivities organised around 'identity politics', and fan cultures keep in touch via globalising technologies, notably in cyberspace, where 'virtual' communities are established (also van Heur, Burgess, this book). In addition, Massey notes that youth cultures are increasingly 'a particular articulation of contacts and influences drawn from a variety of places scattered, according to power relations, fashion and habit, across many different parts of the globe' (1998: 124). These youthful forms of social and cultural relations cut across many of the hierarchies of scale – such as the local, the national and the global – which are held to identify spatial particularities and are organised into 'constellations of temporary coherence' (ibid.: 124–5). This insistence of the permeation of creativity through networks is not to deny that many creative practices mobilise a powerful sense of place, whether in the allotments (Crouch), housing estates (Edensor and Millington), back alleys (Milbourne) and community venues (Bromberg) discussed in this volume. However, the rather more spatially limited notion of the cluster is unable to ascertain the spread and complexity of many more extensively distributed and dynamic creative processes.

Themes

In order to explore the issues identified above, there are four key themes in the book. First, we focus on the ways in which creativity is conceptualised, regulated, enabled and challenged, to highlight the contested cultural values that surround the concept. Second, we examine how we might better understand the spatialities of creativity and move away from the limited concept of the creative cluster and cultural quarter and towards an understanding of the ways in which creativity is constituted within spaces beyond the metropolitan centre (including the suburb, the rural and the network). Third, we will look at the kinds of alternative, non-mainstream scenes in which ethnic, rural, artistic and counter-cultural creativity prospers according to different values and imperatives. Finally, partly in order to escape from instrumental, economistic thinking about creativity, we explore forms of non-economic creativity that emerge in working-class and other communities that ignore orthodoxies about design and fashion to produce economies of generosity and conviviality, providing a critique which advocates the benefits of vernacular creativity.

Part I

Governing and practising creativity

2 Creative spaces and the art of urban living

Graeme Evans

Introduction

This chapter offers a critical perspective on the instrumental use of arts and creative practices for the purposes of urban regeneration, in contrast to vernacular and everyday culture and exchange. Opening with the dialectical perspectives of Raymond Williams ('community culture') and Richard Florida ('creative class'), the move from community arts to social inclusion, and from cultural to creative industries is charted in the context of British urban and cultural policy regimes. Within this discussion, challenges to vernacular creative practices and places are presented, with examples of how culture is treated in flagship developments. The chapter concludes with a discussion of the forms of resistance by artists and others to the commodification of the everyday and the perils of co-optation by the regeneration process.

> [Richard] Florida treats the Toronto Arts scene as a souvenir ... he doesn't get it.
>
> (Anon, Ottawa, 28 April 2008)

> [W]hen Marxists say we are living in a dying culture, and that the masses are ignorant, I have to ask them ... where on earth they have lived. A dying culture and ignorant masses are not what I have known and see.
>
> (Raymond Williams: 1958b)

These two observations could be said to be a world, if not at least an era, apart. Richard Florida, proponent of the 'creative class' (2002), in his latest incarnation at the University of Toronto, and the late Raymond Williams, author of the seminal *Culture and Society* and *Culture is Ordinary*. Williams had been a champion of vernacular, working-class or at least non-elitist cultural expression and experience, but one that was not bound simply by tradition or custom. He thought that introducing change and exposure to new practices over time was a route to cultural development. This incremental, transformative, inclusive approach is in contrast to the imposition of *grands projets* or schemes, and the promotion of the high arts to those with lower 'cultural capital' which have been an enduring feature of instrumental state arts policy and, latterly, arts and social inclusion interventions.

Florida's notion of a 'creative occupation class', on the other hand, claims a relationship between a particular cultural milieu – by no means limited to or even necessarily including artists and creative industry workers – who are not place bound, but, like flexible capital, are footloose and able to be tempted to (re)locate and congregate in cities and areas that provide certain conditions. These include café culture, cycle paths, night-time economies and a creative buzz that together engender clustering of living, working, consuming and inward investment – property, human and financial capital. This creative ecology is thought to underpin the innovation synergies and spillovers most associated with the archetypes of the Silicon Valley and other university-technology powerhouses of Boston (MIT, Harvard), Cambridge (Silicon Fen), Berlin (Eagle Yard), and the exemplars in city regeneration areas of Barcelona (@22), Helsinki (Arabiaranta), London (City Fringe) – to name a few (Evans, 2009).

What these and their emulators have in common is a 'fast policy transfer' tendency (Peck, 2005), many featuring long-term regeneration and redevelopment projects which have been the subject of planning blight and local resistance. Breaking the impasse over these sites and quarters has been justified by using universal rationales – national and global – with a particular effect on local and community culture in terms of the areas and neighbourhoods within which these new creative spaces are being developed. The *creative* class (and underlying innovation-knowledge-science city mantra) is crowding out the *community* (working or ordinary, implicitly 'non-creative') class. This particular form of gentrification is not particularly novel, in view of the now established systemic regeneration effects from property and public realm schemes, and the shift from use value to exchange value of urban space. However, what is different here is that culture and creativity and their spatial and place-making dimensions are being used in arguments in support of the social and community cohesion impacts of the arts *as well as* the more overtly economic development objectives pursued in creative cluster and class policies. Both sets of policies look to produce forms of distinction in particular places within which creativity is to be established. The idea that a creative cluster and 'class' group could be located in a green-field site and housed in a new business park, as with other industries, would be anathema. For the vernacular is an essential backdrop and condition for the new creative quarter, at least to begin with.

This conflation and competition for creative and cultural space is, however, a far cry from the roots of arts and community development practice, which today leaves vernacular and community culture at the margin and faced with joining the creative industries or urban regeneration regimes in pursuing economic or social, rather than cultural, aims. As Garnham also observed, 'there is likely to be a lack of fit, if not direct opposition between policies designed to support [arts/cultural] "excellence" and policies designed to combat social exclusion; the stress of access fits very uneasily with that strand in creative-industry thinking which wishes to reject a hierarchical division of cultural forms and practices' (2001: 458).

In what follows, three sets of cases illustrate the complex relationships between cultural policies and economic cultural developments, and vernacular cultural practices. The first examples represent efforts to provide cultural venues as sites for social inclusion, in the wake of recent policies to combat exclusion. The second group of cases foreground the role of culture in aiming for community cohesion within a context of urban growth and the consequent government 'Sustainable Communities' strategy. The third set of cases document how efforts to promote ethnic arts spaces (and by extension, multiculturalism) have become entangled with the broader shift from cultural to creative industries, as encapsulated in the 'creative city' agenda. In all three sets of cases, emphasis is on flagship developments that are perceived to meet the goals of urban regeneration, but which fall short in providing everyday cultural spaces that address the needs of the local populations in which they are situated.

From community arts to social inclusion

The foundations of community arts practices in the 1970s coincided with the first wave of major youth and structural unemployment, and urban regeneration policy and programme responses in the UK and in Western Europe generally (Evans and Foord, 2000). These practices had some resonance with Williams's democratising community culture in that they were largely place bound, with identified 'communities' engaged (or not) in experimental and compensatory arts activities (Kelly, 1984). Sites included arts and media centres, youth and community centres and housing estates. Arts in education, community radio, artist studios in industrial buildings and *agitprop* theatre were notable cultural responses to the effects of social change during this era. Arts centres themselves have had a particular relationship with the vernacular since they have predominantly been housed in second- and third-hand buildings – from churches, drill halls, factories, to town halls, with over 50 per cent of urban centres located in buildings that were over 100 years old (Hutchison and Forrester, 1987). In 1969, there were 180 projects claiming to be *arts labs*, and from a survey conducted in 1970 there were over 60 designated (i.e. professional) arts centres (Evans, 2001: 90). In a 1986 survey, over 250 arts centres were reported by the Arts Council, but in 1996 only 129 were listed and by 2006 a much-reduced 98 arts centres were included in the latest survey (limited to funded, 'legitimised' centres), with only 64 per cent of these actually calling themselves an 'arts centre' (ACE, 2006). The extent and distribution of community arts centres is therefore understated in official reviews, not least those associated with faith, migrant (e.g. Caribbean, Chinese, Polish 'cultural' centres) and special interest (e.g. art form) groups. The foundation of many arts centres and community arts facilities draws as much from local action as from 'planning', with most established as the result of action by local residents or an arts or community organisation (e.g. school, college) to establish a facility, as well as local authorities seeking to improve local provision or, more recently, to 'regenerate' an area (ACE, 2006).

From the early 1980s the community arts movement and associated sociocultural rationales fell foul of *dirigiste* arts policy – led by 'economic importance' and 'urban renaissance' imperatives – and consequent funding regimes (Hewison, 1995; Pick, 1991), as well as the associated liberalisation of leisure and consumption spaces. It was not really until New Labour's readoption of social inclusion policies that arts and inclusion and 'access' again became a cultural policy imperative. The concept of 'social exclusion' 'had emanated from the brutal housing estates of the Parisian suburbs, to be adopted by the European Union, then the UK government through a newly formed Social Exclusion Unit (SEU). This included policy reviews of the Arts and Exclusion at community and neighbourhood level' (Shaw, 1999). Subsequent interventions, in large part to compensate for the neglect of community and youth cultural activity and resources, included Creative Partnerships (Arts in Schools/Youth programmes) and a decade (1995–2005) of lottery-fuelled capital reinvestment in the cultural infrastructure (arts, heritage, sports), primarily existing arts buildings and organisations (Evans, 1998; 2004).

The new social-cum-cultural policy imperative has also given rise to the development of new facilities in new locations. In contrast to the reuse of industrial buildings for artists, designer-makers and performing and media arts spaces that are commonly found in the post-industrial city of today (Hutton, 2008), new-build cultural facilities have had mixed success and reception. Some have failed within a year or two of opening – Sheffield's National Museum of Pop Music (now the local Student Union centre), the National Centre for Visual Arts Cardiff, Bradford's National Faith Centre, Life Force (a £5m attraction which received 62 visitors in its opening week) – while others struggle to complete and operate such as The Public 'digital media centre', West Bromwich. The original development organisation for this centre, Jubilee Arts, had been forced into administration as cost over-runs (from £40m to £62m) undermined this overly-complex facility, with no clear artistic function. Still, the regional Arts Council claimed that the centre had 'already made a tremendous contribution to transforming West Bromwich, helping to kick-start other long-term regeneration projects that will bring economic, cultural and community benefits to the area' (Luton, 2008). When the digital gallery had to be closed as the computerised exhibits failed, the Arts Council withdrew its outstanding funding less than a year later, leaving the centre's future uncertain.

The local perspective on The Public, below, encapsulates the difficulty faced in developing new arts facilities and 'edutainment', in an area poorly served by mainstream amenities. But clearly the local community has not been engaged or considered in such a top-down, star architecture-driven venture:

> the public is a complete waste of money! Sandwell and Black Country needs better schools, cinemas, theatres, swimming baths, ice rinks etc not a £52million white elephant! I have 2 children aged 12, 9 they rather spend £20 at the cinema watching the latest pixar film which has more artistic merit than all of you clowns who run, and said yes to the doomed project in the first place.
>
> (*Building Design*, 13 June 2008)

Ironically, the supermarket chain Tesco has submitted a joint planning application next to The Public for a mixed-use retail and office development, with cafés/restaurants and leisure facilities – with a local primary school and police station due to relocate to this site. Perhaps the prosaic will be more popular than the prophets of the new media digital experience.

On the other hand, higher-profile art museum conversions such as Tate Modern on London's South Bank and regional galleries such as the Ikon, Birmingham, and the Baltic, Gateshead, have survived, more or less successfully – although heavily reliant upon public subsidy. But while the institutional and national cultural centres and events retain a residual value and importance, despite in many cases their declining popularity and narrow class base (Bunting *et al.* 2008), it is the everyday lived cultural practices and experiences that signify, to borrow Willis's phrase, 'common culture'. As Willis argued:

> the new temples of High Art ... may enjoy some corporate popularity, but as a public spectacle not private passion, as places to be seen rather than to be in. The prestige flagships are in reality no more than aesthetic ironclads heaving against the growing swell of Common Culture. Let's follow the swell.
>
> (1991: 13)

Willis also suggests, less reactively, that some of these mainstream cultural institutions should also be focal points and facilitate partnerships and collaborations with local arts and cultural activities and networks. For example, the development of local libraries and museums through more animated and accessible forms of interpretation; arts in the community, health and education; and the use of interactive technology, could be seen to offer a bridge between the sterile high and popular culture dialectic, and he suggests a more cultural democratic approach, again echoing Williams:

> The recent successes of certain museums and art galleries in appealing to a wide range of people and communicating with new audiences, and the continuing success of many libraries in providing an ever wider range of symbolic materials, rest not upon extending an old idea to new people, but on allowing new people and their informal meanings and communications to colonise ... the institutions
>
> (Willis: 12)

Community venues such as arts centres also serve a dual purpose, including a social role as informal meeting place – not always reflected in audience/user figures: 'around half of all users visit for social reasons, independent of their attendance at, or participation in, arts activity. For most, this social use is occasional, but a core of around 13 per cent of attenders use their arts centre for social purposes on a frequent basis' (ACE, 2006: 49). Conversely, venues such as pubs have played host to regular theatre, comedy and music performance – folk clubs, Sunday jazz, pub rock – including resident companies and early 'arts labs' (Schouvaloff, 1970). In this sense, users adapt and adopt informal cultural spaces and communal venues according to their social and collective needs, not those of curators or arts policy makers. Lefebvre recognised the tension in the term 'user', which had something vague – and vaguely suspect – about it. '"User of what?"

one tends to wonder. The user's space is lived not represented' (or conceived) (1991a: 362).

While much community arts and cultural activity and facilities have been incremental and, to a certain extent, developed organically, the scale of major development projects and population growth together challenge traditional cultural planning and community development approaches (Evans and Foord, 2008). Social inclusion objectives in this scenario have been subsumed into wider sustainable development and communities meta-themes.

Creating cultural opportunities in sustainable communities?

The latest incarnation of state concern for community culture can be seen through the rationales underlying the UK government's Sustainable Communities strategy (ODPM, 2003). In this case, culture is attached to the government's sustainable development, 'quality of life' and economic growth goals, particularly around population and housing expansion in and around major areas such as the Thames Gateway and Milton Keynes South Midlands (MKSM) regions. This has required planning for new and extended settlements at a scale not witnessed since the 1960s and earlier post-war new town developments – with a target of 3 million new dwellings. What these new and changing communities might look like – how their cultural and social aspirations and needs might be reflected in amenities, services and the design of spaces – are therefore questions that have not been posed in living memory and practice, certainly not in a society that is no longer homogeneous or static but which is 'mixed' (socially, tenure-wise, culturally) – with considerable mobility and churn, and both inward and internal migration.

What vernacular actually means in this dynamic context and how arts and culture – and heritage legacies – might be reflected in the urban ecologies that are emerging, is also not clear. This is particularly the case where the community does not yet exist *in situ*, for instance in new urban villages, but also where densifying populations comprise incumbent and new lifestyles and vernaculars, some of which will have originated elsewhere. This is played out in the everyday, for example, through school curricula and extra-curricular activities, celebrations and hol(y)days; through high streets, food, fashion, music and through the use and reuse of buildings and public spaces.

From new town to growth region

The challenges of planning for culture in a context of growth can be illustrated in the case of Milton Keynes (MK), the iconic planned, post-war new town, and in the development of its civic theatre. While no regional arts planning exercise was undertaken for this new town, the potential for a large theatre was highlighted in the development blueprint in the 1970s. In 1985 the MK Development Corporation reported that the creation of a live performance space would be highly desirable and, following a successful bid for National Lottery funding, an award of £20m was made towards the £30m cost of a theatre and gallery. In 1999 the theatre opened: in the words of the Council, 'in addition to bringing a variety of performances to the city, Milton Keynes Theatre provides a focus for the city's already thriving cultural life'.

From another perspective, however, this traditional theatre is felt to lack a certain spirit. In response to an audience question, 'what would you do to make MK a place where arts were a contemporary and necessary experience?', the theatre director Sir Peter Hall said: 'build a smaller theatre for a start'. The present theatre is a dehumanising space. It's well attended because, presumably, there is nothing else that gives you the beginnings of that kind of experience, but it's not a congenial theatre (Hall and Hall, 2006). His namesake the academic planner added,

> I think MK is difficult precisely because it is so completely new. MK central is the most totally created, planned space that we have in this country ... but I think the problem with MK is that it has been too successful. So it does not have any derelict spaces
>
> (ibid., 2006)

The distinction between (artistic) content, the flagship facility and the importance of 'place' – cultural and symbolic – is apparent from these observations. The idea that building a new theatre is necessarily the right type of provision or the complete answer to local cultural provision is obviously questionable (Evans, 2005), particularly given the realities of funding a venue reliant on touring shows and with no in-house production resource. A 'thriving cultural life' may not be the impression that either residents or visitors would have of this 'city'. The Theatres Trust – the national Advisory Body – also makes the point that a town that already has a lyric theatre within 30 minutes' drive is unlikely to need another, but there might well be demand for an arts centre or other small cultural facilities. A strong connectivity with cultural facilities and spaces to learn and exchange demands a local catchment, with the neighbourhood level – including schools, community centres, churches, parks – providing the most regular and highest rates of participation in arts, crafts and group activity (Evans, 2001), underlying the 'power of the everyday' (Lefebvre, 1991b).

This suggests that the planning and provision of cultural amenities and facilities driven by development opportunities and an inappropriate use of 'place-making' can neglect community and cultural needs, and the imperatives of both accessibility and public choice. In the ongoing national survey of cultural participation by the UK culture ministry, *Taking Part* (DCMS, 2007), the key barriers to participation in arts and cultural activity were found to be not only 'access' – location/transport, cost/entry price – but also the relevance ('subjects I am interested in') and quality of cultural activity and events on offer. In short, community and more vernacular culture that reflects the experience and interest of local audiences and participants. Current sentiments suggest that this is an issue in this growth region. Residents in the town of Wellingborough, Northants, in the MKSM growth region, when asked how they felt about opportunities for participation in local decision making mostly disagreed that they had an influence on decisions affecting the local area: access to facilities was a problem (public transport) and between 30 per cent and 48 per cent said they had *never* visited their museums, theatres and concerts halls, while those who did went infrequently (Wellingborough Borough Council, 2007). Satisfaction with theatres, museums and galleries was also lowest in the neighbouring districts, in contrast with more ubiquitous amenities such as libraries and parks and open spaces.

'We're all creative – *now?*'

The cases highlighted above illustrate the instrumentalisation of culture in social and sustainable development policies. However, the recent shift from a focus on 'cultural' to 'creative' industries within UK policy discourse and a corresponding emphasis on the economic value of such activities also has implications for what kinds of cultural activities and spaces are valued. Pratt (2005) notes that this shift reflects a political project that can be traced back to the late 1990s, when centrist 'New Labour' sought to disassociate itself from the left-leaning 'Old Labour' and its support of cultural industries (GLC, 1985). The shift is also linked with the increasing focus on 'information' and 'knowledge' as bases of competitive advantage (Evans, 2009). In 1998, a government-instituted task force on creative industries defined such creative industries as 'activities which have their origin in individual creativity, skill and talent and which have potential for wealth and job creation through the generation and exploitation of intellectual property' (DCMS, 1998: 3), foregrounding the significance of the economic – relative to cultural – aspects and valuation.

Gibson and Klocker (2004) argue that beyond the role creativity can play as a generator of economic activity, another concern of contemporary policies is how creative industries can distinguish or brand places, thereby contributing to urban and regional economic development. Indeed, following the 'creative city' principles advanced by Charles Landry (2000) and by Landry and Bianchini (1995), a number of national and local policy makers have sought to identify the cultural assets that could distinguish a given place and promote those assets that could simultaneously add value economically and reinforce ethical values. In practice, however, this strategy has tended to privilege certain 'creative' activities and a consumption-oriented approach to arts and cultural development and fuelled a copycat creative city movement (Evans, 2009).

Hidden art and rich mix

The application of this broader policy orientation within the UK can be exemplified in the growing popularity of ethnic arts and the development of branded 'multi-cultural' spaces. Within amateur arts activity, this subsector figures prominently today. Table 2.1 presents the findings of a survey of amateur and voluntary arts participation. The scale of engagement includes 50,000 organised groups represented by nearly 6 million members and 3.5 million further volunteers taking part in over 700,000 events attended by 158 million during the year. While amateur dramatics and music are the most popular, 'multi-art', including ethnic and new art forms, makes up the largest and most active group.

Table 2.1 Amateur and voluntary arts, 2006/7 (DCMS, 2008)

Art form	No. of groups	No. of members 000s	No. of extras/ helpers 000s	No. of performance/ events	No. of attendances 000s
Craft	840	28	13	3,000	924
Dance	3,040	128	12	57,000	10,906
Festivals	940	328	395	12,000	3,481
Literature	760	17	11	4,000	191
Media	820	62	12	21,000	1,563
Music	11,220	1,642	643	160,000	39,325
Theatre	5,380	1,113	687	92,000	21,166
Visual arts	1,810	265	‚52	8,000	1,289
Multi-art	24,330	2,339	1,692	353,000	79,789
Total	49,140	5,922	3,517	710,000	158,634

Most creative practice and 'making', such as crafts, amateur and youth art, is still undertaken in vernacular (or everyday) settings, including ethnic and community culture, and in interstitial spaces. From the skateboarders and young graffiti artists outside the concourses and undercrofts of the arts complexes of the South Bank, London, and MACBA, Barcelona, to the 2+ million ballroom dancers that meet regularly in local halls and clubs, even before the advent of TV's *Strictly Come Dancing* (and presumably one reason for its audience success, attracting 9 million viewers each week).

Moreover, in the extreme of 'fringe' cultural display and exchange, locations are also more often to be found on the edge of the city, such as raves in warehouses or fields, which commonly attract audiences from a 50-mile radius, and weekly community markets – selling crafts, antiques, food, clothes and household goods – as in ethnic quarters and in second- and third-world cities, under the shadow of motorway flyovers and football stadia. Dance, music and entertainment acts intermingle with these markets, which regularly draw participants from a wide area of the city and surrounding regions. Cohen argues that 'the most prominent examples of cultural fusion in the arts do not come from global centres but rather from the world's periphery; they represent primarily an attempt at localization of global stylistic trends – the fusion of Western artistic styles or forms with local third or fourth-world cultural elements' (1999: 45). For Werbner (1999), the exchange goes both ways, where migrants from the 'periphery' bring and develop a knowledge of and openness to other cultures that create new hybrid opportunities *within* the metropolitan core. Writing about the British Pakistani community, she argues that this cultural group has engaged in a complex traffic of objects-persons-places-sentiments which has altered the perceptions of 'Britishness' and enabled the creation of a British Pakistani culture (Evans and Foord, 2004).

A new and prescribed attempt to capture the multicultural city in physical form and place – and recognising its absence and marginal position in the past – is seen in two cultural facilities in East London, the Rich Mix Centre and the Institute of International Visual Arts (INIVA). Located in the city fringe area of the borough of Tower Hamlets, which contains several of the most deprived neighbourhoods in the UK, and host to past and recent diasporas from Europe and Asia, these new-build arts centres on former industrial sites aim to be a focal point for local communities, a meeting place for entertainment and cultural education. They also seek to challenge and strive for creative excellence over a range of art forms – working towards a new understanding of British culture. What is being deliberately understated here is the multicultural basis for these ventures, which is manifested, in the case of Rich Mix, in its multi-screen cinema dominated by mainstream Hollywood and occasional Bollywood films, and as home to music training agency Asian Dub Foundation; and in the case of INIVA, visual art/ photography exhibitions of work by 'artists from different cultural backgrounds'. Their location (and funding mix) seeks to play a major role in the regeneration of an area that had already been subjected to office and residential gentrification and development. A visitor to these multicultural arts centres would be surprised by the white faces of the majority of staff – attracting the 'right' skills to operate these facilities from the local community has apparently proved to be difficult. Evidence of ethnic youth and community cultures from these multicultural neighbourhoods is largely absent in 'their' designated arts centres.

These optimistic cultural developments were based on *creative city* principles (Landry, 2000), focusing almost exclusively on creative industries and related retail, hospitality (e.g. curry and balti houses, wine bars, designer retail and galleries), visitor attractions and street markets. Their strained evolution and creation is also indicative of their ambiguous place, situated between mainstream cultural institutions, local regeneration and new cultural practice – but fitting none of these comfortably. At the same time, their multicultural residential neighbourhoods have been neglected by this consumption-led approach, creating a spatial divide with social programmes which promoted training in new media and patronising capacity building, but which ignored the local meaning and memory of place and the cultural knowledge, aspirations and skills of local residents (Evans and Foord, 2004). The rich-mix promise has been reduced to a commodified landscape of street retail and entertainment – a consumption opportunity for adjoining office workers, weekenders and the new urban professional (Shaw, 2007).

Resistance and exclusion

As development encroaches, the vernacular comes face to face with the global, whether masquerading as state or as private interest. While these grand projects and regeneration schemes radically alter the city landscape and locus of mass leisure consumption, they arguably still have less resonance with the everyday places and practices in the residential areas and prosaic functions of the city. Locations that are increasingly reflected in artists' representations now include suburbia, everyday spaces, objects, people, and the artist-as-subject and autobiographer.

Examples include the self-conscious work of Grayson Perry, Sarah Lucas and Cindy Sherman, and in the opening up of studios and workplaces through 'Hidden Art' to the public and vice versa during, for instance, Open House and Open Studios annual tours. As well as celebrating the 'ordinary' (*sic*), this also represents an internalisation of cultural expression and a narcissistic tendency of the contemporary artist and a lack of engagement with the political (and community). This engagement has in some respects shifted to the arena of environmental and social justice movements, rather than the influence of the avant-garde and bohemian radicalism of the past (Wilson, 2003).

> The DiY Culture [*sic*] of squats, anti-roads protests and Reclaim the Streets actions is, among other things, a direct assertion of new cultural possibilities – and a way of living in which culture, art, pleasure would play a central part.
>
> (Edwards, 1999: 2)

Site-based resistance movements are also active in mega-projects such as Poblenou, Barcelona (Kriznik, 2004) and elsewhere, with artist and community intervention in the regeneration process in their own backyards. For example in Sheffield, South Yorkshire, artists have been directly engaged in the process of redevelopment of the city – in gentrifying Devonshire Quarter ('DQ'). Andy Hewitt and Gail Jordan are site-based installation artists with a studio overlooking Devonshire Green near Sheffield city centre. Two projects were commissioned and undertaken by this team, both focused on the DQ area: *Outside Artspace* (2001–2) and *I Fail to Agree* (Hewitt and Jordan, 2003).

In Outside Artspace the artists worked with the city planning department to help develop a vision 'to reinforce the identity of the area and improve land use, transport, urban design, the local economy, housing mix, sustainable living, quality of the environment and community safety' (Hewitt and Jordan, 2003: 26). This area has been associated with youth activity and small businesses serving this market (skateboarding, record shops and cafés) and a growing university student body, due to the development of new halls of residences (Evans and Foord, 2006). During this process West One, a large-scale, eight-storey apartment development was under construction overlooking the only large green space in the city centre. The artists visited the West One showroom to discuss their vision for the development. They said that the council planned to build a bandstand, create a pleasant safe area with CCTV – an image directed at the 'exclusive' apartment market, with the green as a 'front garden' feature for new residents, rather than a community, social and public space. The artists' proposals arising from community consultations included a venue for art projects, exhibitions, film, performance, music events – as part of an annual programme – and youth facilities, including a skateboard park. These proposals were received by the Council and contact with them then stopped – the recommendations were not taken up. Five years later, in the master planning consultation exercise, 'concern was expressed that the Green skate park was not shown on the (new) plans and they had heard that it was being got rid of' (EDAW, 2007). Forms of dialogue and engagement proved to be merely cosmetic in this case, a familiar exercise in co-optation.

Resistance is not confined to local artists, but local communities express their anger at the so-called culture-led regeneration process and housing renewal, through community newspapers such as the *Salford Star*. The city of Salford – 'poor cousin' to its Manchester neighbour – hosts Salford Quays, a 1980s redevelopment and central government-inspired (and -funded) urban regeneration zone, now hosting the Lowry Arts Centre and Imperial War Museum of the North, and soon-to-be-relocated BBC studios at a new Mediacity UK development adjoining Salford University campus (Christophers, 2008). This new cultural quarter is served by an extension to the Manchester Metro Light Rail – but which does not go to Salford town centre to where most local people actually live, including young people who have little or no ownership of the arts complex, from which, not surprisingly, they feel excluded.

In 2006, six local lads ('hoodies') from an East Salford estate were asked by the *Salford Star* to visit the centre to see the Lowry painting exhibition (depicting local factory workers and 'working class people off similar estates'). On a wet Sunday afternoon they entered the building, went up the escalator to the exhibition and walked past the information desk, into the gallery. They were stopped within two minutes of entering this 'free' venue and refused entry. Security was called, but no reasons were given for this by the staff. Another visitor at the time commented: 'basically they were local lads coming in to look at the pictures because they were bored stiff and they were denied access to a facility which we've been told is open to everyone'

Responses in contemporary street 'art' – from the transformation of simple tagging and graffiti to the 'signature' work of Banksy and Christo – represent another approach to the perversion and conversion of mainstream culture, but also the cult of the artist-personality and their marketing and promotion. For instance, the graffiti crew that covered the New York subway trains and led to the mayor's zero tolerance in the 1970s/80s has now gone 'legit', working for advertising firms and department stores in Manhattan on large-scale shop displays and billboard art (Evans, 2007).

A social-cultural market has also developed through trade and fringe events (e.g. Designer's Block, London; Design Mai, Berlin) and interventions, as well as arts and creative activities in 'non-arts' venues. These include housing estates, hospitals, parks, and temporary use of spaces for raves, performance art, student shows, time-based installations and digital 'art', and cultural events offered by new communications technology. This perhaps comes closer to the democratisation first envisaged for the cultural industries that the market and new technology offered small producers and communities (GLC, 1985). However, these new creative spaces have a short shelf life, similar to the way in which alternative and 'creative tourism' spaces fast become commodified and subjected to heritage valorisation (Evans, 2007), a process that Wilson also documents in earlier bohemian quarters (2003). A digital divide also persists, which undermines efforts to widen the distribution of much public edutainment, communication and knowledge – in the UK over 40 per cent of the population do not have broadband access, and of those that do, this is no guarantee of the skills and networks required to move beyond the benign e-mail and e-commerce to more creative applications. In the

Valley of Silicon, 'home' [*sic*] to Google, YouTube, Hewlett Packard *et al.*, these hi-tech global operators based in city fringe industrial parks have little connection (or financial contribution) to their San Jose community – which includes a large Hispanic and Vietnamese resident population – or to the downtown cultural facilities (Evans, 2009). In California as a whole, Latino young people are half as likely to have computer access at home – 36 per cent compared with 77 per cent of US-born non-Latinos. The vernacular spaces of creativity for many communities and young people may therefore continue to rely on traditional places of exchange, including local streets and amenities, rather than on the amorphous possibilities of Web 2.0.

Conclusion

Lee, adapting Bourdieu, refers to the spatial sphere as a 'habitus of location'. He suggests that cities have enduring cultural orientations which exist and function relatively independently of their current populations or of the numerous social processes at any particular time: 'In this sense we can describe a city as having a certain cultural character ... which clearly transcends the popular representations of the populations of certain cities, or that manifestly expressed by a city's public and private institutions' (1997: 132). The latter point is important in any consideration of cultural planning, since attempts by municipal and other political agencies to create or manipulate a city's cultural character are likely to fail, produce pastiche or superficial culture, and even drive out any inherent creative spirit that might exist in the first place.

Flexibility over cultural facilities may also require flexible design and informal spaces, as well as dedicated production and participatory facilities to accommodate local needs over the life cycle, particularly when communities are not yet established or embedded. This might, in turn, offer present and future residents 'the freedom to decide for themselves how they want to use each part, each space'. As Hertzberger goes on to suggest: 'the measure of success is the way that spaces are used, the diversity of activities which they attract, and the opportunities they provide for creative reinterpretation' (1991: 170). This is important, if new and evolving communities (and artists) are to have some input into and ownership of the type and range of cultural amenities required to meet their particular creative aspirations and interests.

Creative spaces also do well to resist the attention of cultural policy makers where this is either instrumental or driven by art-form judgements and hierarchies, and also the perspective of cultural places as heritage 'assets' to be conserved, separate from everyday life. There is clearly a case to be made for both valuing and protecting community culture and spaces of vernacular creativity against the twin effects of cultural commodification and gentrification – not least the type that uses culture as a regenerative tool (Evans, 2005). However, value systems that look to the economic impact of the arts and social impact measurements (identifying contested and vague factors such as 'inclusion' and 'cohesion') as the prime rationales for support miss the point. Externalities may of course arise, but their value lies in their cultural impact, not in being a conduit for crime reduction, health improvement and economic development (Evans, 2005; Matarasso, 1999). An identification and enhancement of everyday cultural practices and their manifestation in formal and informal spaces should therefore remain the focus of contemporary creative space initiatives. In this way, vernacular creative spaces may be better placed to accommodate social dynamics and encompass continuity as well as change over time. As Williams observed, this is likely to be a *Long Revolution* (1961).

3 Creativity by design?

The role of informal spaces in creative production

Norma M. Rantisi and Deborah Leslie

Introduction

The significance of creative industries for cities has been well established in contemporary literature (Scott, 2000; Florida, 2002; Markusen, 2006). Studies cite the role that these industries play in cultural development, neighborhood revitalization and economic development. In recent years, the focus among researchers and policy makers has centered on the economic significance of these industries, both in terms of their direct contribution through the production and distribution of cultural goods, and in terms of the role they play in place-branding a city for tourism and cultural consumption. This discussion has been particularly salient for traditional industrial centers, such as Montreal, which are seeking to reposition themselves and forge new identities within the global economy. Creative industries are often held up as a panacea for deindustrialization and the associated job loss or abandoned manufacturing quarters, but the question of how to promote the viable development of such industries remains an open one. Some commentators express skepticism that a creative economy can be planned at all, suggesting that artistic currents emerge organically. Mommaas (2004: 521), however, sees a potential role for policy in fostering conditions conducive to the growth of cultural industries.

In this chapter, we approach this question by examining the policies and practices that have contributed to the development of a vibrant creative economy in Montreal, Canada. More specifically, we consider how formal governance structures, such as policies put forth by public and nonprofit organizations, are complemented by informal 'sites of regulation' (Lloyd, 2004), such as the physical and social resources of everyday life, in contributing to the city's cultural dynamism and the strength of its design sector.

We draw on 26 interviews with institutions involved in regulating design in Montreal, including key actors in local, provincial and federal government, professional associations, nonprofit organizations, design schools and museums. In addition, the paper is based on over 60 interviews with fashion, industrial and graphic designers in the city. Interviews were open-ended and conversational in style. They ranged between one and two hours in length and were recorded, transcribed and coded according to theme. The research also includes an analysis of government and institutional policy documents relating to design.

Citing evidence from these interviews and policy documents, we argue that top-down, or 'institutionalized' policies are in and of themselves not sufficient to effectively support the development of creative practices, since they overlook the complex dynamics that are implicated in these practices, privileging the commercial dimensions at the expense of the aesthetic and social ones. We illustrate how informal 'sites' correct for these limitations and allow for more sustainable forms of creative production. We trace the links between the city's industrial heritage and its rise as a 'cultural metropolis', with a focus on contemporary challenges and opportunities for cultivating a creative economy.

Montreal provides an illuminating window into policies and practices that can nurture creative industries. Within the last several years, it has garnered international attention for its success in this area (Stolarick and Florida, 2006). It was designated a UNESCO 'design capital' in 2006 and is presently home to the International Design Alliance, the Centre for Canadian Architecture and the Cirque du Soleil. The city is also a center for fashion and has a flourishing performing arts and independent music scene. Indeed, 'between 1971 and 1991, census data show that Montreal had a larger percentage of its workforce in the arts and cultural industries than in any other Canadian metropolitan area' (Germain and Rose, 2000). Although more recent data indicate that Toronto has caught up, Montreal has 90,000 employees in cultural industries, representing over 5 per cent of the city's employment (City of Montreal, 2005a). In terms of design alone, the city ranks sixth in North America for total employment, following New York, Boston, Toronto, Chicago and Los Angeles (Design Industry Advisory Committee, 2004). According to the City of Montreal, there has been a 40 per cent increase in the number of designers in the city in the last ten years (City of Montreal, 2006).

The chapter is divided into three sections, and we begin by reviewing the spatial dimensions of cultural industries. We foreground the need for both formal and informal forms of support. We then go on to highlight the origins of cultural policy in Montreal, and the formal institutions that have emerged for regulating culture and design, identifying the rationalities underpinning these programs and their limitations. In the third section we discuss how Montreal designers' ability to create rests on informal everyday spaces and communities.

Spaces of creativity

The interest in creativity within popular discourse and policy circles is tied up with the rise of the knowledge economy and the growing emphasis on new ideas and meanings rather than physical production as the basis of competitive advantage for goods and services (Lash and Urry, 1994; Scott, 2000; Florida, 2002). And while creativity exists across the economy, when applied specifically to the production of cultural goods and services (as opposed to high technology, for example), it assumes a distinct attribute. It represents not only the production of new knowledge but, more importantly, a particular kind – symbolic knowledge. In contrast to scientific or engineering ('functional') knowledge, which relies on information processing, this knowledge requires a deep understanding of the habits and norms of 'everyday culture' and an ability to interpret and communicate cultural trends in a unique way (Asheim *et al.*, 2007).

Indeed, most contemporary scholars highlight the symbolic – relative to utilitarian – qualities of cultural products as their defining feature (see O'Connor, 1999; Thorsby, 2001; Scott, 2001), and an emphasis on these qualities has three further implications for understanding how creativity 'works' within cultural industries. First, cultural industries, like other knowledge economy sectors, must contend with market volatility, short-lived product cycles and the need to continually innovate. However, for other industries, product obsolescence can be accommodated in business plans and market research. For cultural industries, aesthetic judgments operate alongside profit as key business goals and the ephemeral nature of aesthetic tastes renders these industries even more vulnerable to risk and uncertainty (Banks *et al.*, 2000). As a consequence, these industries require resources and networks to balance commercial objectives with aesthetic ones and to manage the risks and failures that are intrinsic to aesthetic experimentation (Frey, 1999; Banks *et al.*, 2000; Castaner and Campos, 2002).

Second, and following from the first point, creativity is a process that occurs through an interdependent system of activities, ranging from conception to production to consumption. These activities involve both cultural actors (e.g. creators or artists) and economic actors (e.g. producers, distributors, retailers) and the flows of knowledge between them (Hirsch, 1972; Becker, 1982; Caves, 2000). Whereas cultural actors have the critical task of developing a concept, economic actors are instrumental in translating the concept into a marketable good or service. The role of the latter can range from establishing the parameters (e.g. the market trends) within which creators can operate, to expanding the market for a product through new distribution channels or marketing initiatives.

Scholars such as Scott (2000) and Molotch (2003) highlight yet a third characteristic – the spatial dimension of creativity. They contend that such activity is likely to occur in cities, where one can find a concentration of specialized services or support institutions and a density of cultural actors. This concentration permits timely access to inputs and information, thereby mitigating the risks associated with cultural production. In addition, the social and cultural diversity of cities can inspire creators to think outside of established conventions, exposing them to divergent practices within their creative field (e.g. fashion) as well as those outside of their field (e.g. music, architecture). And for older industrial cities, in particular, the material fabric (older manufacturing lofts, industrial relics) is said to provide an added source of creative stimuli (Markus, 1994; Bain, 2003; Hutton, 2006).

What are the implications of these attributes for the regulation of cultural industries? The attributes highlighted above underscore the complex nature of cultural production. They suggest that creators embed themselves within a socio-spatial context that can give rise to the symbolic and material resources needed to balance commercial with aesthetic objectives. In practice, once a creator is implanted within a socio-spatial setting, the acquisition (or 'discovery') of such resources often occurs serendipitously (Currid, 2007a). Neither market mechanisms nor hierarchical modes of governance can effectively regulate such a process (Santagata, 2004); and both threaten to 'crowd out' aesthetic experimentation by privileging commercial objectives and/or dictating the nature and terms of production. In an analysis of cultural industry policy, Frey (1999) argues that governments are not well positioned to regulate cultural activities, since they are not cognizant of the specific logics or needs that underlie such activities, particularly the social or cultural dimensions (see also Currid, 2007a; and 2007b). Moreover, Frey (1999) finds that governments are not, by definition, less risk averse or commercially oriented than corporations. Recent analyses, including several of the chapters in this book (e.g. chapters by Bromberg, Evans, Miles and Van Heur) corroborate this view, noting that municipal governments are increasingly adopting entrepreneurial strategies in the face of budget constraints and that such strategies include an instrumentalization of culture (see also Leslie and Rantisi, 2006). This has translated into direct grants or subsidies, primarily aimed at promoting success cases.

Rather than emphasizing direct intervention, a host of studies draw attention to the informal sites of regulation that can shape practices in cultural industries (Mommaas, 2004; Lloyd, 2004; Markusen, 2006; Currid, 2007a). More particularly, scholars note that cultural producers are more likely to privilege aesthetic relative to commercial attributes when provided with resources that socialize risks and motivate experimentation, especially those that give rise to diverse and complementary sets of cultural practices. Such resources include affordable live-and-work spaces, and a dense concentration of diverse land uses, including the presence of what Lloyd (2004) terms 'third spaces,' i.e. unregulated spaces between home and work – cafes, restaurants and bars. These spaces serve as less formalized spaces of inspiration, where cultural producers can hang out, exchange ideas and 'just be' (Markusen, 2006).

Thus, with regards to policy, what is needed is what Frey (1999) refers to as 'crowding in' measures, forms of indirect support that can create a context in which risks are embraced and failure is accepted. Although Frey tends to adopt a more conservative viewpoint and is suspicious of any intervention by government (as compared with trade associations), other scholars, such as Mommaas (2004) and Zukin and Costa (2004), see a potential role for public institutions in nurturing such a context, particularly with regards to the regulation of the spaces and land uses that underpin creative encounters. A common ground for all these scholars is that 'crowding in' measures can counteract market mechanisms and a contemporary policy thrust that emphasizes commercial objectives.

In the sections that follow, we examine the rise of formal policies that support cultural industries and design in Montreal and illustrate how these top-down approaches are primarily aimed at the commercialization of products. We further show how the informal attributes of the city – the everyday spaces in which producers work and interact – provide an alternative and complementary means of regulation.

Institutionalizing creativity: the origins of design policy in Montreal

Compared to other cities in North America, in Montreal a strong policy framework has emerged to support culture and design. This policy imperative reflects the rise of nationalist sentiments in the province of Quebec in the 1970s and the view that culture and the arts could be a means for preserving a distinct Quebecois identity. By 1992, this view took the form of a Cultural Policy which accorded a significant role for government intervention in fostering and supporting cultural industries and institutions. This policy had a particular focus on Montreal as the cultural metropolis of Quebec (*La Politique Culturelle du Québec*, 1992).

Economic restructuring in Montreal since the 1970s has also prompted an interest in design. Like other cities, Montreal faced a process of deindustrialization and decline in mature industries, such as garment manufacturing (Polese and Shearmur, 2004). Faced with heavy job losses, government officials became convinced of the need for a new economic development strategy, and in 1985 a federal committee was established to study the city's economy. The committee's report, known as the Picard Report, identified cultural industries and design as essential elements in Montreal's transition from low-end manufacturing (Picard and de Cortret, 1986).

The convergence of these trends and the policy directions to which they gave rise have led to a concerted effort by municipal and provincial governments and nonprofit institutions to forge a real (and imagined) association between Montreal and design. They have strived to do this through two means: the marketing of Montreal as a design city, and the promotion of greater design intensity in Montreal products and landscapes. Here, we analyze three key institutions that constitute part of the contemporary policy context for governing design in the city: a municipal program, a local design institute and a provincial body to support cultural enterprises. This survey of institutions is not exhaustive, as there are other institutions (e.g. trade shows) that play a role; however, it is illustrative of the basic tenets that have come to define how design is regulated.

Foremost among these institutions is a series of initiatives on the part of the City of Montreal itself. Following from suggestions made in the Picard Report, the municipal government mounted a successful design competition from 1996 to 2006, Commerce Design Montreal. The competition gave awards for the innovative interior design of restaurants, shops and other commercial buildings in the city. Encouraging the use of professional interior designers, this initiative was designed to bolster the international profile of the city and to brand the city as an 'up-and-coming design metropolis' (http://ville.montreal.qc.ca). The idea was to construct a 'ripple effect among neighboring businesses, so as to convince other businesses of the benefits of design for their bottom line, and exert a strategic influence on the revitalization of commercial streets' (http://ville.montreal.qc.ca). A further goal was to foster the development of cultural capital – that would also attach itself to products being manufactured in the city (Interview, Design Commissioner). The competition was later exported to other cities around the world – including Brussels, Marseille, Lyon and Saint-Etienne – illustrating the innovativeness of the city's policy framework. The competition is still run today by a local design group, Créativité Montréal.

Convinced that design could have tremendous benefits for a city's branding and growth, the city also created Design Montreal in 2006 to broaden its sphere of action to include all design disciplines. The mandate of this new unit is to augment Montreal's status as a 'city of design':

> The quality of design, architecture and urban development contributes directly to the excellence of urban life, Montrealers' pride in their city, and visitors' enjoyment. Montreal has demonstrated its know-how and creativity on a number of occasions ... but design quality must now become the city's signature, in its own urban development initiatives and be visible in all boroughs.
> (http://ville.montreal.qc.ca/portal/page)

Design now forms a central component of all municipal policies and the city is sponsoring an economic development strategy that will support design innovation in public spaces, as well as in planning and architecture. The city is also considering the possibilities for establishing a procurement policy that would encourage the use of professional Quebec designers in city projects.

A second institution is the Institute of Design Montreal (IDM). Launched in 1993 by one of the key architects of the Picard Report, this nonprofit organization has a mandate to promote industrial, graphic, interior and fashion design, as well as architecture, landscape architecture, urban planning and new media (Interview, IDM official). IDM receives funding from several levels of government and runs a plethora of programs, including design competitions, forums and initiatives to foster linkages between designers and manufacturers. As in the case with the city, a primary emphasis of this organization is to 'promote design as an *economic value*' (IDM, 2001: 1, emphasis added) – a value that could be harnessed through greater awareness and through an increasing commercialization and volume production of designed goods. Incentives are provided to designers and potential clients (business investors, manufacturers) to realize this value.

A third institution that forms part of the design policy nexus is SODEC (Société de Développement des Entreprises Culturelles) – a public institution that is a product of the 1992 Cultural Policy and is overseen by the Ministry of Culture and Communications in Quebec. The aim of this organization is to support the economic viability of cultural enterprises in the province. Bridging the worlds of art and business, the agency provides grants, loans, export assistance and investment to cultural businesses to enhance the commercialization and distribution of their creations (Interview, SODEC official). SODEC also sponsors internships and design exhibitions that facilitate networking among business investors, firms and designers. As with the IDM, the emphasis here is on the economic value, and to be eligible for support from the organization, designers must establish that they already have a proven track record and a business plan.

While collectively the programs outlined above form an innovative policy nexus to support design in Montreal, we highlight a number of limitations associated with these initiatives. First, there is a focus on the economic dimensions of design and on the commercialization of a concept. For example, there is a strong emphasis on providing export assistance, marketing support or bringing a design to market. While these strategies are crucial, they tend to emphasize the distribution process. Such sites often overlook the spaces and communities needed for incubating design talent in the first place and conceiving a product.

Second, these institutions are formal in nature. The selection processes for forums and competitions often privilege well-known and accredited designers with established track records (Interviews, designers). Emerging designers, designers trained in other creative fields, or designers whose work challenges the boundaries of design tend to be marginalized in awards and assistance programs (Interviews). This can discourage cutting-edge work and unorthodox forms of talent, effectively 'crowding out' aesthetic experimentation (Frey, 1999). As noted earlier, such modes of governance tend to be risk averse.

Third, the formal institutions discussed here operate according to a 'top-down' logic, often seeking to direct behavior. Official institutions play an important role in convening design forums; they typically seek representation from designers, and many of their programs have educational value for their participants. However, the power relations embedded within such sites are often hierarchical. Workshops and competitions are often structured to generate consensus around an already defined agenda. Commerce Design Montreal, IDM and SODEC are designed to foster new governmentalities – cultivating new forms of entrepreneurial behavior and subjectivities (Leslie and Rantisi, 2006; Foucault, 1991). In this way, participation in official programs serves to incorporate rather than to empower participants.

While formal institutional initiatives are important to fostering a vibrant design economy in Montreal, they tend to be commercially oriented, hierarchical and prescriptive. Initiatives of a more formal nature therefore need to be supplemented with policies designed to stimulate more informal sites of practice. We suggest that informal and everyday spaces – or what we call vernacular spaces – play a critical role in generating, maintaining and shaping creative production. In particular, we argue that accessible, diverse, open and affordable spaces are critical to the nurturing of creative talent and the production of symbolic knowledge associated with cultural industries. In the following section, we discuss some of the informal spaces that support creative production in Montreal.

Enabling creative production: everyday and vernacular spaces of support

The role of cultural diversity in artistic innovation

Ideally, informal spaces are accessible to everyone, places where difference is encountered and negotiated (Young, 1990). Open spaces afford the possibility of identifying with different groups and opinions, offering opportunities for inspiration and enrichment. There is an active and emergent quality to these spaces that contrasts with the highly programmed nature of formalized sites discussed above. Everyday spaces are spaces of doing, of production. Cultural and socio-economic diversity can support the production, distribution and consumption processes by providing opportunities to create and test products for multiple markets and language groups.

Montreal is one of the most diverse cities in North America (Atkinson, 2007). This diversity is a key strategic advantage:

> Francophone talent may be less inclined than Anglophone talent to be 'drained' away from Montreal to English-speaking cities. Also it can be argued that Montreal has the potential to become a Mecca for businesses seeking mobile and cosmopolitan key workers for an internationally oriented economy, because of its exceptionally high rate of bilingualism, its increasingly multiethnic school population and growing numbers of young people whose first home language is neither French nor English ...
>
> (Germain and Rose, 2000: 142)

For Germain and Rose (2000: 127), there is, not surprisingly, a clear link between the city's unique cultural linguistic base and its large and diversified cultural artistic centre.

Specifically, for outsiders, Montreal's bilingual and multicultural character means that the city is more open to informal interactions; newcomers who may speak a foreign language can immediately feel included and can exchange their experiences and practices. For locals, openness to diverse experiences and practices can facilitate a departure from the established ways of doing things:

> We have all these different cultures that are enriching us tremendously. This is why I say Montreal could be a hugely artistic city if it is well supported. Yes, even the design in Montreal could become definitely leading, worldwide, because of the mixture of cultures. The disadvantage of being a maestro in Italy doing bags is the knowledge that they have. It becomes their prison. The fact that we don't have this knowledge, as I said it in the past, gives us definitely the possibility of doing much more without even knowing it. Yes, I would say today, I have a huge advantage being in Montreal because I'm not limited to a fifty- or one-hundred-year world way of doing something.
>
> (Interview, fashion designer)

Cultural diversity can also mediate global–local articulations in the city. Global forms of knowledge are increasingly significant in stimulating creativity (Malmberg and Power, 2005); but as Donald and Blay-Palmer (2006) note, these networks are more likely to stimulate unconventional thinking when they flow through a vibrant multicultural city with the resources to synthesize and adapt global trends in novel ways:

> as a cultural experience, it's very rich because you always ... have to adapt yourself to [something] different. There is quite a big multiethnic aspect in Montreal, so that is contributing a lot to Montrealers' curiosity. The way we will eat, the way we will buy clothes or our relationships with the arts, with language. I think all these aspects make Montreal a unique place and the

fundamental important aspect in the background of a creator … But I would say the intangible aspect of this is may be … to be able to go maybe easily towards another culture. Which is not that different, but still, the way your brain has to work in order to express yourself in another language. It makes you do the exercise to communicate a kind of global, or universal …

(Interview, industrial designer)

For another designer, multicultural diversity provides an escape from the homogeneity of design trends, infusing work with ideas from diverse sources:

It can go from ethnic restaurants … It can be … crazy bars where authentic people gather … I think the influence is more subversive than if you're in contact with design spaces because it easily falls into trends, and, okay, 'this is hot so we should do this'. If you want to be the leader in the matter of creation, you really have to turn yourself, to turn towards what's happening inside you as someone who lives in, let's say, in your time, and facing frustration, facing anger and … being in contact with different cultures. So I think these kinds of things really are stimulating in the creative way of speaking.

(Interview, industrial designer)

The diversity of Montreal – particularly in terms of its bilingualism and immigrant cultures – and continuous encounters with that diversity – thus allows one to break out of established conventions, to be exposed to new ideas and to test those ideas in multiple markets.

The role of economic diversity

Not only is it important for designers to engage with people from all walks of life and cultures, but recent studies suggest an important role for economic diversity in providing exposure to divergent practices. Montreal is one of the most diverse cities in North America in terms of industrial structure. And this is a quality that is clearly exploited by designers, many of whom cite the importance of spill-across linkages – the diffusion of knowledge from one industry to another. Stolarick and Florida (2006) suggest that spill-across linkages are particularly important because they supplement spillover effects, which occur between firms within an industry. Following Jacobs (1969), they in fact suggest that knowledge transfers across industries and sectors are the most important source of innovation (Stolarick and Florida, 2006: 1800). For many designers, these spill-across effects contribute to the development of new ideas and allow for cross-marketing. One fashion designer, for example, discusses the inspiration she gets from local artists she encounters in the city:

I'm looking what they've been doing in the silk printing and prints, so it helps to give a direction … for example the first [collection], I wanted it to be really strong and austere and East European and cold and very light, so that was the direction.

(Interview, fashion designer)

The same designer also cites the potential production and marketing benefits of such spill-acrosses:

> Every collection, I'm working with a different artist so it's like they work with a print and I tell them look the direction of the line for this season and we sort of work together. And they make a print for the collection. So every collection ... is a 'silographie' ... a silk printed on the clothes in production. For this collection, the prints were done in collaboration with graphic and other fashion designers ... for my next spring collection it's another painter friend of mine who's going to do the prints. So every project, it's a collaboration and I try to put their name always so people know who did the prints. Collaboration is always interesting; it opens new doors.
>
> (ibid.)

Moreover, in addition to these advantages, an economy that includes a diverse range of independent actors (producers, suppliers or distributors) can afford creatives the opportunity to take part in developing their own communities and spaces, often outside the dominant economy. To quote one fashion designer:

> In my case, where I'm a small company, all my pictures I've done with a photographer who is a friend. What we did is I paid for the cost and he uses the pictures for his portfolio. No one really got paid for any of it. The models were his friends and I gave them bags. Or, even for the last fashion show, I paid her in clothes. We do a lot of exchanges. I work with a graphic designer and in exchange she'll take some clothes. I also like to work with people who are at the same stage in the sense that we can all learn together. They're open to what it is I am. Sometimes if you work with people who are too professional, they're working at a different speed. They're not patient. When you're still at the beginning, where you're kind of exploring your whole creative process, you want to work with people who are exploring as well, or have less limitations.
>
> (Interview, fashioner designer)

Thus, a diverse range of actors enables the possibility of exchange not mediated by money, but rather by direct encounters. This type of exchange was noted by several of the designers whom we interviewed and such exchanges would often occur informally, on the basis of social connections, and would lead to the development of other contacts (Interviews).

Informal sites of practice: the neighborhood and 'third space' encounters.

In addition to cultural and economic diversity, spatial diversity is also cited as a key ingredient for enabling creative networks and practices. Within Montreal, zoning for mixed land use – residential alongside retail and commerce – and mixed income has led to the development of self-contained neighborhoods with distinct identities (e.g. Mile End, Plateau, St Henri). Since the 1990s, for example, the municipal government has sought to revitalize inner-city neighborhoods by promoting territorial bounded 'villages' through a range of initiatives. Such initiatives include the revitalization of commercial streets and restrictions placed on the development of inner-city shopping centers, the preservation of existing housing stock, investments in the maintenance of green space and community facilities, and support for local neighborhood cultural centers (Germain and Rose, 2000). As a consequence, these neighborhoods offer a community feel, and strong ties within neighborhoods can be important because they promote an overall sense of well-being that is conducive to survival in a volatile industry. In particular, feeling comfortable and at ease in one's surroundings is beneficial for designers, who often experience high levels of risk associated with short-lived product cycles. Even if they are mundane or run down, it is the quality of social ties in neighborhoods that is important to designers, rather than the presence of highly aestheticized landscapes.

Within neighborhoods, the mix of land uses gives rise to local 'third spaces' such as cultural centers, restaurants and cafes, which provide multiple opportunities for casual interaction between people from different walks of life. Third spaces are hybrid spaces which transgress and subvert dualistic categories (Soja, 1996; Bhabha, 1990). Such sites are fragmented, fractured, incomplete and uncertain; they are spaces where identities are constantly negotiated through encounters with others. As Bhabha (1990) argues, third spaces are not concerned with identities but rather with identification – the process of identifying with and through others. As such, third spaces are sites of struggles over meaning and representation; they open up possibilities for new structures of authority and new political initiatives. Such locations are particularly important for fostering creativity and new ideas.

Montreal is known for its vibrant public spaces and outdoor life. Germain and Rose (2000: 79) suggest that renewed interest in public sociability can be attributed to the prevalence of such places, especially when they spill over onto public streets, and argue that the social diversity of the city makes these places open and welcoming. As Markusen (2006) notes, designers value spaces as places to engage with their peers and with other members of the creative community. The local cafes, cultural centers and well-known artist hang-outs provide outlets where designers can learn from one another and encourage and critique each other's work (Markusen, 2006; Lloyd, 2006). They can serve as sites of exhibition, for example, as in the case where restaurants sponsor regular fashion events or cafes display the creations of local artists (Interviews).

Such informal settings can also provide supportive ties that help to connect designers to resources. As one designer puts it,

> There are lots of restaurants. There are lots of cafes. It is a bit of the same feeling as Paris ... It is a neighborhood where you see a lot of people in the streets ... You get a real connection with people in the street, [and feel] that you are not really isolated in your work. And I don't really want to feel that when I go to work, I'm really going to work. I just want to feel that I am just going to take a coffee. It's a psychological mechanism to not get stressed.
>
> (Interview, graphic designer)

'Third spaces' are often flexibly organized, offering the possibility to 'drop in' at any time. Their informal nature enables designers to piece together information and contacts from diverse sources.

Affordable spaces

A final element of the everyday experience in Montreal that contributes to creative production is its affordability. In Montreal, studio and living space is relatively inexpensive. This relates in part to the role of government with regard to housing (Germain and Rose, 2000: 163). In particular, the City of Montreal has made substantial resources available to nonprofit groups to recycle or renovate older buildings for low-cost housing. The city has also maintained an affordable housing supply through rent control and public housing initiatives.

Low rents are a significant part of the innovation equation. They provide designers with a modicum of security, making them less reliant on conventional clients and more likely to take risks and to privilege artistic integrity:

> Montreal is quite an exciting place to live for a creator because, first, the rents are quite cheap. So the quality of living we have here is allowing us to do a lot of activities. Going to theatre, to movies, eating in restaurants because you don't have to put necessarily all your money into rent. You can also rent for a respectable amount, a design studio.
>
> (Interview, graphic designer)

Thus, affordable spaces translate into lower barriers to entry, whereby designers can focus on their creative pursuits, including the exploration of local cultural stimuli that will inspire new design ideas.

Not only do lower rents provide the conditions for innovating in terms of design, they also free designers to engage in other forms of artistic practice. A majority of designers and cultural workers are interdisciplinary and are active in more than one art form (Molotch, 2003: 179). Here, a graphic designer describes how low rents and supportive artist buildings enabled her to start her own art gallery on the side:

> The cost of living here and the kind of cultural importance they put on art and making things and community and being involved, regardless of culture, regardless of age. It's so important for fostering that environment where you can really go out and open a gallery. And you might meet your goal and fail in it after a year and people will still commend you ... I like the healthy balance right now that I have in Montreal between making money at what I do and not having to compromise my aesthetic because the support is there for trying.
>
> (Interview, graphic designer)

By lowering the costs of failure, low rents provide designers with the conditions for working across fields and experimenting with activities that they may not otherwise undertake.

Conclusion

An examination of the role that informal spaces can play in regulating design challenges the notion that policy should be exclusively formulated in a top-down and targeted manner, with the goal of creating a design culture anew. Indeed, like Frey (1999), our research suggests that artists and other cultural sectors require 'crowding in' measures, or indirect forms of support, that serve to counteract market mechanisms and facilitate the creation of an environment in which risks and failures are acknowledged and even encouraged (see also Mommaas, 2004).

This is not to suggest that policy does not have a place in nurturing creative spaces. In particular, we find that it is social policies of a more universal nature that are particularly helpful for designers, including those that foster housing affordability, income redistribution, social inclusion and the development of open and accessible public spaces. For example, rent control policies, immigrant settlement and integration policies and initiatives that create festivals celebrating Montreal's diverse cultural heritage help to foster an overall landscape of acceptance and difference that offsets some of the risks associated with the production of symbolic knowledge or insecure employment.

While the formal policies targeted at the design sector are useful in that they build a more general awareness of the value of design and support the economic imperatives of commercialization and marketing, these need to be complemented with other kinds of initiatives that can nurture the creative and aesthetic dimensions of cultural production. As the case of Montreal illustrates, a range of more general social equity and quality of life initiatives, which enhance already existing attributes, can enable everyday sites of creative practice.

4   Art goes AWOL

Malcolm Miles

Introduction

In this chapter I introduce current art in the UK which constitutes a dissident vernacularism. By this I mean art which departs from, and actively contests, cultural policy and which arises at a grassroots level. Cultural policy in the UK has tended, since the 1980s, to revolve around claims for art's expediency in dealing with a range of social and economic problems. This emphasis on art's utility replaces an insistence on quality derived from the autonomy of form claimed for modern art, and proclaimed by bodies such as the Arts Council since the late 1940s. The emphasis on art's socio-economic utility reflects a move, in the 1980s, from arts administration for the public good to arts management based on a business case for investment. In following global agendas this policy differs, also, from the local agendas and participation characteristic of community arts. The policy is most often seen in culturally led urban redevelopment. But if redevelopment schemes look to the requirements of competition in global markets, dissident art contests the principles and the efficacy of such top-down policy.

Indeed, while the new dissidence remains professional art, in face of market-driven tendencies epitomised by Young British Art since the 1980s, a growing number of art graduates have formed groups or collectives to work outside the criteria of both cultural policy and the art market. Policy and the market are, of course, linked. The role of flagship cultural institutions such as Tate Modern in London or the Guggenheim in Bilbao, for example, is key to the symbolic economies of the cities which host them. In the UK, within a policy of arm's-length arts management, which includes the setting of funding levels for Tate Modern, Tate Britain, Tate St Ives and Tate Liverpool, the state does not intervene in aesthetics, but inevitably influences the policy direction of institutions. Those institutions also play a significant role – in what they exhibit – in a consensus among curators, dealers, critics, collectors and a few of the more successful artists as to what constitutes contemporary art's mainstream. To emergent artists' groups, that consensus excludes politicised art and marginalises art made outside a small number of metropolitan cities.

The cases of current but non-contemporary art on which I focus, then, are the work of artists who have qualified in higher arts education and make at least part of their living from art practice (while many derive income from multiple sources, including teaching and residencies in non-gallery settings). This work constitutes a third category, neither the contemporary art of museums nor the more obviously vernacular art of local art societies and evening classes in life drawing, photography and various crafts. The new dissidence thus has a vernacularism which is grassroots, but grounded in high art while in some cases intervening in local agendas. It can be objected that the history of modern art is a history of departures which have generally been integrated into the mainstream. Marcel Duchamp's once-provocative ready-mades are available to collectors in limited editions, and Andy Warhol's images of celebrity always were. Perhaps today's dissident art will be similarly absorbed. The mainstream is now able, as I argue below, to incorporate almost anything. But dissident art has its own means of dissemination in digital or web-based communications, through radical cultural networks and informal alignments. It reacts to the threat of assimilation in the mainstream by largely ignoring arts institutions and using arts funding only when this does not prevent criticality (including of the arts funding system).

Asking whether dissident-vernacular cultural production can challenge the orthodoxies of art and urban development, I outline the growth of radical planning (as an informative parallel development), discuss art's expediency and consider cases of dissident vernacularism in London, Liverpool and Plymouth. Some cases reference global issues, others relate to local cultures. All, however, reconfigure cultural work in terms of engagement local to the site of production and the milieu of producers.

Points of departure

I begin by contextualising the issues in terms of a parallel tendency in radical planning. Radical planning has retained criticality since the experiments of advocacy planning in the 1960s (Davidoff, 1965), in which planners promoted local interests to planning authorities. Like dissident art, radical planning takes a grassroots approach, inasmuch as planners spend time with members of communities, but not so much to represent them as to help them articulate and represent agendas for themselves. If policy is strategic, concerned with overarching imperatives, radical planning and dissident art are tactical and discursive. The latter aspect differentiates professionals educated in specific (and changing) discourses from lay people (though members of community groups can be professionals in other fields) and from amateurs. But if expertise was the means by which planning authorities excluded dwellers from decision making or arts institutions maintained aesthetic hegemonies, in radical planning and dissident art the knowledge of professionals is set beside, not above, that of dwellers.

Radical planning was a departure from the rational comprehensive planning model developed in Chicago in the inter-war period. That orthodoxy sought to order urban growth through technical expertise, replacing the vicissitudes of political control with what it claimed as an objective, scientific mechanism for dealing with rapid growth. For Chicago urbanists, whether sociologists or planners, the city was a laboratory to be understood using biological models (Short, 1996: 179–80). For E.W. Burgess, for example, waves of inward migration produced effects 'analogous to the anabolic and katabolic processes of metabolism in the body' (Burgess, 2003: 160). Burgess drew on biology as a means to naturalise his urban theory, putting it outside history and ideology. Interventions based on technical expertise were immune to contestation by localised interest groups whose tacit, local knowledges were regarded as inferior. The iconic representation of this approach is his concentric ring diagram (Burgess, 2003) based on Chicago's central business district and bands of transitional neighbourhoods and suburbs. Subsequent generations of planners may or may not have tried to assimilate the development of other cities to the diagram (Savage and Warde, 1993: 16), but the centrality of a business district is more than a matter of geometry, anticipating today's commercialisation of inner-city spaces. Doreen Massey argues that the distanced view of the bird's-eye city plan, of which the concentric ring diagram is a further reduction, privileges visuality as the sense offering mastery (Massey, 1994: 222–4). Reducing the spaces of everyday life to a white ground, the conventional plan allows decisions to be made as if that ground is void. Roads can thus be driven through residential neighbourhoods in pursuit of progress, regardless of the human consequences (Berman, 1983).

In contrast to the reductive view of the conventional city plan, Peter Marcuse (2002) proposes a layered city of overlapping, intersecting zones of business, production and dwelling. Within each, class and ethnicity produce edges. New York, then, is several overlaid cities: a city of wealthy residential areas; a gentrified city; a tenement city; and an abandoned, post-industrial city of waste lots. Overlaid on this, again, are sites of corporate control, service industries, production, unskilled work and semi-legal or illegal activities. The effect is a mosaic-like urban texture denoting a city of contested spaces of political and ideological agency in which victims are produced by economic change and policy. How they are treated 'depends on who they are and how they react', while their status determines 'the policies a state will follow' (Marcuse, 2002: 109). Similarly for Nancy Fraser, status differentials are reinforced by design, and markets are instrumental in differentiating status by 'bending pre-existing patterns of cultural value to capitalist purposes' (Fraser, 2003: 58).

In contrast, Michel de Certeau views the practices of everyday life as the resistant waywardness of dwellers whose 'tactics interrupt the state's or market's strategies of order, as evidence of a social imagination …' (de Certeau, 1984: 41). It might be thought that art contributes to the shaping of such an imagination by lending visibility to its outcomes. The difficulty is that the tactics of everyday life are (like the judgements of amateur art) experiential, not the result of intellectual transactions or discourses, while cultural production is negotiated consciously, even oppositionally, as the representation of a set of conditions and awarenesses. In culture there is no raw, authentic revelation, only its mediation in a verbal, visual or other sensory language. In cultures, in an anthropological sense, the acts of daily life are negotiated too, but tacitly as expressions of shared values (and what is possible or allowed) which are not consciously predetermined but are generally accepted. This is not to say that vernacular architecture or craft goods are devoid of aesthetic qualities – they can be regarded as beautiful – but that these qualities emerge in reception of a materiality which is not, as such, a discourse.

Art does not have that option. What it has, though, is the possibility to intervene in discourses and the categories of discourses. Public and site-specific art failed to do this. Meanwhile, grassroots opposition in art took the form of a knowing use of visual culture's vocabulary – as in a poster campaign in London's Docklands during its redevelopment, in which The Art of Change worked with local networks to tell stories of a labour history erased in the area's redevelopment (Dunn and Leeson, 1993). This is not, however, an equivalent of de Certeau's 'everyday act', such as resistant walking, which operates outside institutional structures. Radical planning and dissident art work in relation to (but interrupt) institutional frame-works, inevitably informed by what they seek to revise from a critical or incidental position. Walking, in contrast, remains simply walking, unless it is incorporated in a cultural discourse such as that of Situationism in Paris in the 1960s. There, wasting time by drifting around the city was regarded by Situationists as a meaningful waste of time in a society aligned to productivity. But that was art.

Radical planning or a cultural turn?

Leonie Sandercock writes in *Towards Cosmopolis* of the radical planner as working on the edge of the profession to utilise planning knowledge for the benefit of specific, local publics. Citing insights gained by Lisa Peattie and Allan Heskin in the non-affluent world, she writes that the radical planner

> immerses herself in the community, hangs out with them, helps them with research and preparation of documents, advises on how to deal with bureaucracies, but never does things *for* the community, always *with* them. The identity of the radical planner in these works is that of a person who has, essentially, gone AWOL from the profession, has crossed over ... to work in opposition to the state and corporate economy.
>
> (Sandercock, 1998: 100)

This, as argued above, requires knowledge but makes that knowledge multi-centred. Sandercock notes that to go too far can lead to professional oblivion, and cites John Friedmann's argument for critical distance from, rather than absorption in, localised struggles (Sandercock, 1998: 100, citing Friedmann, 1987: 392). Distance alienates social actors who shape local agendas, but being too far inside the situation makes the planner a marginal figure lacking influence with political establishments. Sandercock concludes that the role of the state is crucial for radical planning. The state does not engage in radical planning, but it is 'misleading to think that radical planning can do without the state' (Sandercock, 1998: 101).

Today, the state tends towards deregulation. This lessens its role in relation to social welfare – though recent events in the global financial crisis may lead to a revision of the state's role in relation to that of market forces. Indeed, intervention is historically associated with crisis. An example of intervention on a local scale, still playing to a political agenda of social order, is the Conservative government's decision to fund a garden festival in Liverpool in 1984, following unrest in Toxteth – one of the city's most deprived districts. The festival was intended to bring jobs and prosperity to a city in decline, assuming a relation between unrest and economic conditions. But it was also a spectacle aimed at changing the city's image, informing a new symbolic economy.

Garden festivals were held in Gateshead, Glasgow, Stoke-on-Trent and Ebbw Vale in succeeding years, each one bedecked with a quota of public sculpture. The garden festivals were an early form of culturally led regeneration aimed at levering inward investment while showing the state's concern for social problems The siting of flagship cultural institutions in inner-city districts – like Tate Modern in Southwark, one of London's poorest boroughs at the time – is a more ambitious and expensive version of the policy. The outcome is a boom in leisure and consumption sites, arts buildings, public spaces and designated cultural quarters in cities across the post-industrial West.

The latter constitute an aestheticisation of space in a period of immaterial production. The cultural turn in urban development ignores the possibilities for a more democratic determination of future scenarios that are offered by radical planning, since culture – despite its association in classical thought with goodness and truth – becomes the emblem of redevelopment which is often socially divisive. Just as the nineteenth-century public realm excluded women (Wilson, 1991), aestheticised spaces tend to exclude publics unable to buy into cosmopolitan affluence. Monica Degen observes that, from empirical research in the Castlefield quarter of Manchester, 'residents feel their sense of place is disrupted as they feel invaded by activities and sensescapes produced by outsiders' (Degen, 2008: 176). And in redevelopment in Barcelona's red light district, now El Raval, she sees urban cleansing: 'by stimulating more civilized activities in the area, less space will be available for undesirable activities' (Degen, 2008: 181). This extends a history in which modernist architects and planners saw inner-city streets as disorderly, to be replaced by mono-functional spaces. Hence the acts of everyday life which intermingled in inner-city streets are separated out and allocated separate spaces (for living, washing, parking, leisure and so forth) in a neat but dehumanised vision of spatial ordering standing for a specific social ordering (Robbins, 1996). For Jamie Gough, Aram Eisenschitz and Andrew McCulloch, such pursuit of civic order is 'intended to improve the quality of services and labour power available to production, to defuse antagonisms in employment, to strengthen social order, and thus to re-legitimate capitalism' (Gough, Eisenschitz and McCulloch, 2006: 186).

The question, then, is to what extent art's expediency is also a means of social ordering, or re-legitimation of the market. In the 1870s, in France, Germany, Britain and North America, a proliferation of public monuments and statues was designed to buttress an otherwise questionable sense of national identity (Michalski, 1998: 13–76). Today, the arts are used to address any social or urban issue (Yúdice, 2003: 9–39). An underlying strategy of assimilation seems to be one element in this, as shown by the summer 2008 project, *Street Art*, at Tate Modern.

Art's expediency

Tate Modern has, in many ways, been successful, its visitor figures exceeding all expectations. Yet it is seen, too, as replacing the bourgeois cultural elitism of earlier types of museum with creative commerce (Leslie, 2001). But, as Esther Leslie goes on to argue, its public face must remain cultured, not explicitly marketised. *Street Art* seems to epitomise a strategy for visible democratic inclusion while being at the same time a means to add a new category of product to the art market (validated by Tate as a major institution). The exterior walls were painted by graffiti artists from Bologna, New York, Paris, Sao Paulo and Barcelona – each in their distinctive style. Stating that *Street Art* was the first major museum display of graffiti in London, the show's sponsor invited visitors to 'see street art in its natural habitat' on a walking tour of Hackney (*Metro*, 14 August, 2008: 44). *Street Art* reads graffiti as a global art movement which 'celebrates the dynamic form of urban visual art' (ibid.: 45). Graffiti was regarded in New York in the 1980s as 'garbage, pollution, obscenity, an epidemic, a disease ...' (Cresswell, 1996: 37), and remains anti-social behaviour for local authorities, who erase it from public walls. Tate's use of it as a sign of urban vibrancy replicates the conscription of street styles by fashion, on which Sharon Zukin writes:

> Styles that develop on the street are cycled through mass media ... where, divorced from their social context, they become images of cool. On urban billboards advertising designer perfumes or jeans, they are recycled to the streets ... the beachheads of designer stores ... are fiercely parodied for the 'props' of fashion-conscious teenagers in inner city ghettos. The cacophony of demands for justice is translated into a coherent demand for jeans.
>
> (Zukin, 1995: 9)

Street Art shows that grassroots resistance – in a form more striking than de Certeau's walking – can be institutionalised. Chicago sociologist Louis Wirth wrote of the city as where the behaviour of the 'fluid masses' was 'unpredictable' (Wirth, 2003: 101). If graffiti is a sign of unpredictability, Tate Modern has re-categorised it as art so as to erase its otherness and render it safe. A more local effort to accommodate graffiti was less successful. A local authority in Cornwall unveiled a graffiti wall in October 2008. The first inscription stated: 'I paid my tax + all I got was this lousy wall' (Taylor, 2008).

Dissident art

Sandercock writes that in societies of 'many different cultures, each of which has different values and practices, and not all of which are entirely comprehensible or acceptable to each other, conflicts are inevitable' (Sandercock, 2006: 40). Conflict is inadmissible in the post-industrial state of seamless consumerism, or in its cultural enterprises. An increasing number of artists have reacted against this cultural bromide, and I turn now to cases of dissident art in London, Liverpool and Plymouth.

Platform is a London-based arts group whose core members, Jane Trowell, James Marriott and Dan Gretton, have devised projects on the city's neglected rivers, the meshing of power and trade in the city's history, and the role of the oil industry in environmental and social injustice. In 1996 and 1997, Platform produced two issues of a spoof newspaper, *Ignite*, with news of the oil industry absent from the mainstream press, distributed free at mainline stations. Trowell takes guided walks in central London (Figure 4.1); and Marriott, with researcher Greg Muttitt, takes narrated walks in the financial district, contributes to publications on oil (Muttitt and Marriott, 2002), and holds alternative oil industry annual meetings to which analysts, journalists and campaigners as well as artists are invited. Platform's events involve small-group discussion and aim to seed ideas through prolonged contact. Platform's website states, 'the walk … [is] an important form for public space work … as a research tool, as a ritual, as performance, as intervention, as a political tool, and as a tool for sharing insights and information' (www.platformlondon.org.uk/fitc.asp). But Platform is not activist in the sense of, say, anti-roads protestors in the 1990s, and does not use direct action. It has had contact with the anti-roads group Reclaim The Streets (Wall, 1999: 87–8, 127–8; Jordan, 2002: 60–5, 69–72), who produced another spoof paper, *Evading Standards*, using Platform's facilities in 1997. But Platform disowned this. However, Marriott has since become more relaxed about categories, not minding that Platform was described in a press report as a human rights group (Mcalister, 2008).

Platform's work may constitute a reoccupation of a public sphere of social and civic determination. For Fraser, the bourgeois public sphere was exclusionary (Fraser, 1997: 69–98). Inasmuch as it is aligned to the nineteenth-century public realm of grandiose spaces, statues and public institutions, exclusion is demonstrable. That public sphere was, too, constructed as a polarity to a private sphere of property ownership and family, the rights of which were protected against state intrusion. Hannah Arendt sees the private sphere, however, in a perpetual twilight, precisely because to live privately is 'to be deprived of the reality that comes from being seen and heard by others' (Arendt, 1958: 58). For Zygmunt Bauman, that sphere of free interaction has been for the most part evacuated now, as privatisation encroaches on public space. Bauman sees the task of critical theory as reclaiming this metaphorical site for emancipation. Noting the substitution of quasi-personal issues for public debate in reality television, Bauman concludes: 'The search for an alternative life in common must start from the examination of life-politics alternatives' (Bauman, 2000: 52). Platform's work may contribute, fusing global issues and personal concerns, 'wary of the dangers of the parachute project' in which artists are perceived as 'landing … from outer space, staying for a while, stirring up emotions … airlifting out' (www.platformlondon.org.uk/education/asp).

Similarly located between art and activism, in contact with Platform, the Liverpool-based Institute for the Art and Practice of Dissent at Home (IAPDH) aims to 'resist and intervene in the reproduction of repressive ideologies' in the context of the Capital of Culture programme (Liverpool-08). IAPDH reads Liverpool-08 as 'an excuse for retail, construction and property industries to make good profits … at the expense of local people and grassroots culture' (private e-mail communication, 2008). The Institute is 'a space for dissenting … run by dissenters for dissenters', set up by doctoral graduates Gary Anderson and Lena Simic with their children Neal, Gabriel and Sid, in their flat, and funded by 10 per cent of their teaching salaries (ibid.). *Miss Julie in Utopia 21062008*, was 'a free

Figure 4.1 A guided walk in central London

event with limited capacity' on 21 June 2008, re-performing Strindberg's play *Miss Julie* 'with a feminist, anti-capitalist make-over' (ibid.), with guest artist Cathy Butterworth. Participants wrapped the artist's council in a red banner. As the drama began, a revolutionary soundtrack was played. A party followed. Anderson and Simic described the event as 'history revisited! ... the Internationale sung! ... Strindberg undone! A journey from 1888 via 1968 to 2008. Another world made possible. This was the revolution staged at home' (ibid.).

The utopian intent of *Miss Julie in Utopia* contrasts with the tone of main-stream artist Nathan Coley's neon project for Liverpool-08, a neon sign sited at Tate Liverpool stating: 'There will be no miracles here'. Coley reflects the city's enduring economic decline, putting Liverpool's year as Capital of Culture in its place. But the contrast is not a matter of positive utopia with negative actuality, but between different tactics in the face of de-industrialisation and the perceived fracturing of the social fabric. At first reading, Coley's slogan may suggest a sense of becoming in which everything is in perpetual flux; hence no sudden, revelatory transformation occurs. But this overloads the work with significance. Despite the reference to the 'Internationale' – fanciful but not ironic – I do not think Anderson and Simic are less aware of flux. But their use of rhizomic and incremental dissemination is a more critical and continuing acceptance that there is no magic solution to the city's ills. Anderson and Simic counter the conditions of post-industrial dereliction in a scenario of widening differentials of wealth and status by reclaiming histories of radical culture and action. Despite the demise of state socialism after 1989, and the failure of the utopian project of modern plan-ning and design, Anderson and Simic do not abandon hope for a better world – referencing global anti-capitalism in their adaptation of its slogan, 'Another world is possible'. The question is how.

For Jeremy Gilbert, working in cultural studies, a sense of becoming is an alter-native to the tactic proposed by radical philosopher Ernesto Laclau (2005) 'to identify personal points of antagonism between currently hegemonic projects and widely shared sets of popular aspiration and assumption' (Gilbert, 2008: 187). Yet, by locating dissent in the home and linking the home to radical cultural networks, Anderson and Simic act out that scenario, merging Laclau's idea of making contra-dictions visible into a view of social change as a process of becoming in which the means are the ends. There are further issues: how cultural work has agency; and how dissent operates when housed in domestic space rather than party or campaigning structures. Drawing on Laclau's idea of coalition building as the characteristic of alternative politics, Gilbert suggests a new, open form of political community:

> any counter-hegemonic politics ... must try to plot vectors of potential becoming which do not merely subsume one set of identities and potential points into another. What must be sought are both affective resonances and potential points of symbolic and discursive articulation between existing groups ... but also processes of exploration and invention which might allow unpredictable new becomings to emerge.
>
> (Gilbert, 2008: 189)

For Gilbert, Reclaim The Streets made affective and resonant links while drawing out the conflict of a residual English ecologism with policies favouring car ownership in a 'rhizomatic network of localised, molecular connections' (Gilbert, 2008: 191). I read the work of Platform and IAPDH as informed by a concept of a fluid, open and networked political sphere. Platform builds links to campaigning groups; IAPDH, equally localised in what it addresses, adapts histories of revolution to a vernacular project. For both, networking is the key tactic.

In the first of what is intended as a series of artists' retreats – spaces for collective reflective action – in September 2008, Anderson and Simic brought together artists engaged in politicised and polemical work, including Trowell from Platform, guest artist Ange Taggart, with Abi Lake, Caroline Wilson, Lorena Rivero, Janice Harding and Steve Higginson. In a park at the highest point in Everton a banner was stretched between trees stating: 'The Concept of Culture is Deeply Reactionary'. The aim was to map Liverpool-08 – or 'Capitalism of Culture' – by walking the city's (literally) high and low points, discussing the subjectivities of capitalism. A box containing £500 (the project budget) was placed in the centre of the group as participants introduced themselves. After a 10-minute silence, each took a different route, gathering again in Edge Hill. Then, a reading of Guattari's essay 'Culture: A Reactionary Concept?' led into planning for phase two. Among events on 20 September, Anderson and his son Neal attended a political meeting to discuss recent arrests for leaflet distribution, and made a sign reading 'I am not allowed to give you this leaflet' (e-mail communication, 2008).

IAPDH is unusual in many ways, not least its purposeful merging of dissidence and domesticity. Other events and exhibitions in Liverpool also offer an oppositional stance to Liverpool-08. The Royal Standard is an artist-run space in a redundant factory in Everton, keeping the name of the pub in Toxteth in which it began. Its first group exhibition in the new space, *Navigator*, includes a film by David Ball, *If I Found One I'd Have to Call It Bob*. This weaves a narrative from the loss of a consignment of 28,000 yellow rubber ducks from a container ship hit by a storm in January 1992. The film shows ducks fancifully washed up on the banks of the Mersey, contemplating the brown tide and looking out to sea. Others, the soundtrack narrates, are journeying for ever around the globe, carried by invisible ocean currents. Outside, in another project, a brown shipping container houses part of Oliver Walker's work *Mr Democracy*, in which the absence of a written constitution for the UK – one of three countries globally to have such a lack – is addressed by having one made. Following patterns of global production, the manufacture was outsourced to the East China University of Politics and Law. Meanwhile, 1,000 plastic dolls were produced in a factory, each with a digital sound recital of the constitution. The dolls were sent (again by sea container) to Liverpool, and the progress of the container ship carrying them was mapped in real time on a lap-top screen inside the container at The Royal Standard. Walker writes that the new constitution 'offers an insight into the notion of democracy' which the West so keenly exports, but 'from an alternative perspective', referencing 'the hidden processes of mass production and global trade, so integral to our relationship to China' (Walker, 2008). The constitution states 'Citizens enjoy freedom of speech through various means' (article 19);

and 'All citizens whose rights are infringed … are entitled to require legal relief' (article 29) (www.mrdemocracy.org). This contradicts the prevention of leaflet distribution cited above, but Liverpool is a city whose publics still protest. On the way to The Royal Standard to see *Mr Democracy*, I saw a group of local parents protesting against the threatened closure of an inner-city school.

Citing the repetitiveness of regeneration scenarios in the city since the garden festival in 1984, artists Penny Whitehead and Daniel Simpkins, at The Royal Standard, produced a free newspaper, *Future Visions of History,* for distribution in walks through the city on 30 October, 2008. Commissioned by the independent photography gallery Open Eye, *Future Visions* drew a parallel between the garden festival and Liverpool-08. As the artists say, the garden festival was intended to revitalise the local economy, attracting three million visitors, yet had little long-term benefit for adjacent neighbourhoods, which are still subject to multiple deprivation: 'The festival site was intended to endure as a functioning public space', but it has since been sold to private developers for high-rise, up-market housing and is derelict (Simpkins, 2008). The derelict site stands as a reminder of the irrelevance of such spectacles to the city's social needs. Capital of Culture is similarly a veneer for a future scenario based on shopping (Figure 4.2). Simpkins and Whitehead write,

> Twenty-five years on, the focus has shifted from leisure to culture but the purpose remains to rejuvenate the city by creating a global image worthy of private investment and an infrastructure in which free-market economies can thrive. … As early-career artists based in Liverpool, first hand experience of Capital of Culture has been disquieting. Not only has this year seen the … privatisation of city-centre space … but the redistribution of European funding in the run-up to 2008 has encouraged further institutionalisation of the local art scene, with less visible organisations and projects taking considerable funding cuts to support high-profile and spectacular projects, as the focus has shifted from local sustainability to international marketability.
>
> (E-mail to the author, 30 October 2008)

Future Visions draws on local and invited artists, writers and academics (including myself) to add a critical edge to discussion of the city's future prospects, and the gap between redevelopment of property and regeneration of local economies and communities. Whitehead and Simpkins see their project as spreading ideas 'like a virus throughout the streets of the city' and an 'antidote to the city's hegemonic literature' (Simpkins, 2008). Again, the connections are molecular and the tactics rhizomatic; and *Future Visions* draws, too, on the contradiction of an English pastoral myth reproduced in garden festivals and material life at street level. Simpkins and Whitehead observe a rise in grassroots creativity and anti-Liverpool-08 culture, despite lack of resources (but not of resourcefulness). They read *Future Visions* as their 'attempt to gatecrash the Capital of Culture party' (ibid.).

Artists such as those cited above negotiate a route between art's structures of funding and validation, the autonomy of artist-run projects and spaces for which a range of means of support in cash and in kind are used, and political awareness.

Figure 4.2 Daniel Simpkin: 'Shopping Cultural', 2008

Dissent in Liverpool became public when the Capital of Culture programme was announced, exacerbated by conflict within the city council (Topping, 2008). The new mall is a provocation, its glossy commercialism and disguise as a series of streets (in an open-air version of the classic North American model) conjuring a mythicised democracy of consumption. As Bauman writes, mall shoppers find 'what they zealously, yet in vain, seek outside: the comforting feeling of belonging ...' (Bauman, 2000: 99). But it is community without effort or vigilance, while the powers that manage this effortless togetherness are 'masterforgers' trading in semblances (Bauman, 2000: 100).

Like Sandercock, Bauman emphasises a need to live with rather than erase difference. But there are issues, also, of lending visibility to the cultures and knowledges of specific publics. I want to end by looking at a project in Plymouth, *Reciting the City* by Mike Lawson-Smith, in July 2008. To enlist memories and place-associations from local people, Lawson-Smith set up a studio at Plymouth Art Centre in which to show a selection of short films from the South West Film and Television Archive. Spectators became co-producers of the project, in that their viewing of the films they chose was itself filmed, with sound, for a parallel projection (on two screens) of the film, and the film of its reception, in the gallery at a later date. Lawson-Smith writes, '*Reciting the City* explores how the historical moving image prompts our memories (and expectations) of the geographical shape of the City of Plymouth ... [through] personal histories, moments of recollection and accounts of passed-on stories ...' (Lawson-Smith, 2008). The films were made by enthusiasts and semi-professionals; each lasted between five and fifteen minutes, covering subjects from fishing to local events and industry (including the Naval Dockyard) and reconstruction after heavy wartime bombing. Participants, usually in small groups or pairs, could view as few or as many films as they wished, sitting on a couch while the artist sat nearby. They were encouraged to make verbal responses, but with no script or prompting. Participants were self-selecting, but came from diverse backgrounds and ranged in age from the elderly to young parents with children. Some spent a short time. One couple watched all the films (and ate all the biscuits). In the edited gallery showing of the sessions, Lawson-Smith selected footage to ensure all participants were included, watching at least one film, and that all the archive films were used – to give a total of four hours' gallery viewing. The Art Centre's visitor figures indicated the popularity of the project as compared with other exhibitions, with a fair proportion of repeat visits during a month of screening.

Reciting the City makes no overt protest and is unrelated to specific redevelopment plans, but draws out aspects of its everyday cultures which, lent the visibility of film, could be a resource for planners, urban designers, politicians and future art projects. While the work of Platform and IAPDH confronts global issues, these are obliquely referenced in *Reciting the City*, as participants recall, for instance, the appearance of particular sites prior to redevelopment, and associate places with life events. Lawson-Smith comments:

> Although the project could well be applied to any place, the city of Plymouth ... has dramatically changed physically and socio-economically over the pre- and post-war periods. In the light of the regeneration of the city and the corporate repackaging of its past as a way of generating commercial opportunities, the project is intended to ... offer the chance of revealing otherwise hidden views and histories.
>
> (Lawson-Smith, 2008)

At the same time, the project departs from art-world conventions of single authorship: participants co-produce the work, mainly in groups; their responses to the films they watched form a new oral/visual history layered over the history of enthusiasts' films which were elements of the visual culture of the day when they were made.

Those films were not movies, in most cases viewed only by spectators connected to their production. As amateur art is the art of today, documentary films and club productions constitute a widespread vernacular which warrants visibility. For another example, factory film clubs in Eastern Europe prior to 1989 made films outside state requirements, evolving their own adaptations of themes, from biography and romance to westerns and crime stories (Lewandowska and Cummings, 2004). In the conditions of post-industrial urban redevelopment, and the commercialisation of history, it might be argued that work which brings vernacular dialogues into public hearing is another form of dissident art, or at least a passive resistance to cultural hegemony.

Stretching the category

Networking and collaborative production characterise cultural dissidence. My picture is sketchy but the cases cited raise a number of issues. First, there is a need to radically revise the construction of meaning in cultural forms in plural, democratic and open ways. Artists have sought to do this since the 1960s in projects in local, social and community settings, but the need is also to ask how authorship is reconfigured in the search for a more directly democratic society. A radical vernacularism would involve artists, like radical planners, handing over the means of production to participating groups and individuals whose tacit and intellectual knowledges are given equal status to those of professionals.

Second, concepts of place, site and community require revision, too. Art critic Lucy Lippard writes that 'While the notion of place in art has become more broadly interesting to artists ... it is applied so generally as to become locally meaningless' (Lippard, 1997: 277). In face of global consumerism and its impact on localities, often via the cultural industries, individuals and groups network globally. This lends them a sense of belonging to a community of purpose. Community may now have little link to geographical site, being constituted by shared interest or concern, so that locality becomes, in these conditions, a site of contested social determination. It may or may not be aligned with place in a geographical sense. What emerges is that the modes of communication used – described as rhizomic by Platform – are not unlike older, everyday modes – such as the oral reporting of events. Today, social networking and the web add to the available vocabularies as well as the means, but the outcome is still an implicit model of society not as a node of power surrounded by margins of service, but as multi-centred and permanently multivalent. In oral history, as an event is reported from one person to another, adaptation occurs – usually according to already-known stories. Eye-witness accounts are prone to this: for instance, an account of the toppling of the Vendôme Column during the Paris Commune of 1871 speaks of the head of the statue of Napoleon which had adorned the column being broken off and lying in a ditch. A photograph taken the next day shows the statue intact (Miles, 2004: 17, n9). But rolling into a ditch is what traditionally happens to the heads of tyrants, and the point of the narrative is that Napoleon was a tyrant. At this point, oral vernacular history becomes, in its terms, a knowing discourse. As well as a need to revise categories of site, then, there may be a need to revise definitions of discourse and what separates professional from amateur cultural production. Locally made films, after all, tell stories, and the stories derive from what was known by those who made them.

Third, the work cited above departs from the conventions of art as aesthetic objects validated by arts institutions. This raises the difficulty of a history of departures. If this art is outside the mainstream, and is not the art of today (as amateur production), it requires another name – for which I juxtapose the terms critical, dissident and vernacular. Such work operates between art and everyday life, to intervene, to make contradictions visible, to listen to non-privileged voices. Platform confronts a global industry; IAPDH challenges the hegemony of cultural programmes and bureaucracies complicit in the privatisation of urban spaces. In

Mr Democracy, Walker shows the imbrication of global commerce and constitutional niceties taken for granted in the West. *Reciting the City* and *Future Visions of History* create visibility for contested ownership of the processes of urban development. Such work assists participation in imagining future urban scenarios other than those of top-down planning or market forces.

Yet if everyone is an active user of a creative imagination, envisioning future social and economic possibilities – an artist in Joseph Beuys' terms – the category of art becomes stretched to a point at which it is no longer useful. That may not matter. Such work enacts a creative tension between art and everyday life, culture and cultures, departing from the teleology by which the questions raised and issues contested in intermediate cultural spaces would otherwise be closed. Going AWOL means keeping the issues open in work which may be ephemeral but which for participants may yet be transformative.

Part II
Decentring creativity

5 Creativ* suburbs

Cultural 'popcorn' pioneering in multi-purpose spaces

Alison Bain

Introduction

In the wake of work by Charles Landry, Peter Hall, and Richard Florida, 'creative' and 'creativity' have become two of the most widely used words in the English language. While these two words have been inclusively applied by bureaucrats, urban planners, and scholars alike to an eclectic range of activities, objects, individuals, and places at different spatial scales, from the body to the neighbourhood to the city, they have rarely been applied to the suburb. Frequently maligned as 'boring, uniform, isolated, domestic [places] … full of identical people doing identical things' (Pile, 1999: 29), the caricatured sprawling suburb with its six-lane roads, deserted bus stops, industrial parks, strip malls, big box parking centres, prayer palaces, and garage-fronted McMansions would seem to be the very antithesis of creativity. Yet, ironically, the central values of suburban life – space, privacy, independence, freedom, and personal control – are many of the same qualities that are as appreciated by socio-culturally diverse suburbanites as they are by cultural workers who form the core of Richard Florida's creative class (Wyszomirski, 2000).

This chapter discusses vernacular suburban creativity through a detailed empirical examination of one inner suburb of Toronto, Canada – Etobicoke. This case study is part of a larger SSHRC-funded research program that examines cultural production on the periphery of Toronto and Vancouver in the suburbs of Mississauga, North Vancouver, and Surrey. The suburban landscape of Etobicoke is interpreted through an analysis of in-depth, semi-structured interviews with nineteen cultural workers, cultural planners, and arts administrators who were asked to reflect on the everyday suburban environments in which they live and work. Particular attention is directed to the perceived functional advantages of suburban cultural production in an effort to present suburbs as the new vernacular creative edge of city regions. This chapter argues that a shortage of formal cultural facilities in the suburbs necessitates an alternative reading of the built fabric that reveals the value of improvisational space and multi-purpose buildings. The discussion opens with a consideration of how creativity has variously been interpreted by scholars and practitioners of the cultural economy.

Thinking critically about creativity and the creative city

In a fascinating book titled *On Creativity: Interviews Exploring the Process*, John Tusa (2004), the managing director of London's Barbican Centre, explores how the creative process in the art world is a continuous, uncalculated, and instinctive continuum of innovation and evolution. In artistic practice, creativity involves 'fresh ideas, new problems, and the possibility of new solutions. It is the reverse of repetition, of a formulaic approach to work' (Tusa, 2004: 11). In its original usage, Tusa (ibid.: 6) explains that creativity meant 'the exceptional act of imaginative discovery and expression in an art form'. Creativity as the messy and difficult experiential process of creation by artists was surrounded with an aura of mystery and treated as a 'prized feature of the human mind' (Boden, 1994: 1). It was regarded as 'one of the highest level performances and accomplishments to which humankind can aspire' (Taylor, 1988: 99). It was variously admired as 'insightful, wise, divine, inspirational [and] productive' (Lavie *et al.*, 1993: 5). And then Richard Florida came along and eliminated all of the mystery!

Creativity became an ordinary human ability that could occur everyday and everywhere. Florida applied the concept of creativity to a broad 'creative class' of professionals in business, law, engineering, science, and healthcare who supposedly share similar approaches to complex problem solving, a common work ethos that values individuality, difference, and merit, and are attracted to the same kinds of urban places that offer the three Ts (technology, talent, and tolerance). Not only is everyone creative, but every city now has the potential to tap into the creativity of its residents as well. Where is the difference and the uniqueness that once made creativity a distinctive rather than a common attribute of individuals? When you consider the work of skilled practitioners, it often relies on the qualities of sensitivity and intuition – qualities that are difficult, if not impossible to quantify or standardize. As Arlene Goldbard (2006: 101) goes on to argue in her book *New Creative Community: The Art of Cultural Development*, 'those who focus too closely on "models," "replicability" and "best practices" tend to produce dull work, lacking depth and heart'. By extension, when understandings of what constitutes urban creativity are standardized through cultural policies, cultural plans, and creative city networks, it could be argued that the creative process is necessarily limited, if not lost altogether.

The metaphor of the 'creative city' is supposedly an effort to think about cities differently and to develop a framework for problem solving. In its ideal form, the creative city would foster urban innovation and self-reliance through interdisciplinary and holistic thinking and partnerships based on inclusivity, accountability, and justice. Solutions would break with formula and tradition and be perceptive, bold, and far-sighted. However, in a neo-liberal era of municipal financial constraint where discourses of urban entrepreneurialism and competitiveness dominate and business elites hold the reins of power in public–private partnerships, solutions are rarely socially progressive, risk taking, or daringly original. In such an urban entrepreneurial framework, market forces are readily accommodated while social justice objectives are all too easily abandoned (Miles, 2005; Kingwell, 2008). Moreover, the grassroots urban spaces needed to incubate vernacular artistic communities, creative practices, and solutions are rapidly disappearing, erased through the process of gentrification and eclipsed by the drive of private patrons to import international starchitects to create globally recognized iconic buildings that will appeal to tourists and the creative class and add spectacle value to the urban fabric. But creative people and projects need to be based somewhere.

A creative city needs land and buildings that are affordable and in close proximity to other cultural amenities. Cheap spaces reduce financial risk and encourage experimentation. Free zones allow the flexibility for new arrangements and for improvisation. As Robert Elmes, the director of Galapagos Art Space in New York City has recently said: 'Ideas flourish when there are opportunities to realize the ideas, and that's done in spaces. If you can't afford the space to incubate the idea, you don't get to have the idea come to fruition' (Houpt, 2008: R1). Space is central to the creative and to the improvisational process.

The necessity of improvisational space

Improvisation in the world of jazz is often interpreted as the act of spontaneous creation based on unconscious intuitions (Nachmanovitch, 1990). It can be described as 'the act of exploring what lies at the edge of the unknown as it unfolds in real time without a scripted plan' (Bain, 2003: 312). Two scientists at Johns Hopkins University who specialize in the neuroscience of music made the print media news with the results of their study on where creativity comes from (Neergaard, 2008). They studied the brain scans of six professional jazz pianists when they played from memory and when they improvised on a plastic keyboard inside an MRI, and concluded that creativity uses the same brain chemistry as dreaming. The brain circuitry supporting self-expression and sensory awareness is turned on, while inhibition is turned off.

This scientific research on creativity relates to my earlier work (Bain, 2003) on the importance of 'improvisational space' at the neighbourhood scale for stimulating the creative practice of urban visual artists. I argued that artists often choose to live and work in neighbourhoods in which 'diversity, flexibility, playfulness and even outrageousness are both encouraged and accepted. The opportunity for artists to explore, to look, to listen and to shift the boundaries with which they experience the world is integral to their creative practice ... In this precarious and unsettling creative state on the edge of discovery and uncertainty, a delicate balance is usually needed between freedom and order, play and structure within the urban spaces of artistic production' (Bain, 2003: 312). The opportunities for finding and securing improvisational space in cities – space that is changeable, malleable, and affordable, that encourages spontaneous and intuitive activities, and that supports different work arrangements – are rapidly diminishing. It used to be that such space was found in low-income, working-class, inner-city neighbourhoods. But the condominium boom has intensified residential development in the downtown core to meet the consumptive and experiential practices of middle- and upper-class 'fauxhemians' (with their corporate jobs, trust funds, and bohemian lifestyles) (Kingwell, 2008) and in the process displaced artists and creative industries. The neighbourhoods that remain affordable and can act as incubators for culture and entrepreneurship are often now located in the inner and outer suburbs, not the central city.

The suburbs, I argue, are the new and under-appreciated creative spaces in city regions. Why did the late, great jazz pianist Oscar Peterson move to the Toronto suburb of Mississauga in the 1970s? One journalistic interpretation is that 'In the suburbs, Mr. Peterson found a new, more open society. Although it was largely white, Mississauga seemed more amenable to change, if only because it lacked the crushing social history of downtown Toronto … It was a case of not being rejected … In Mississauga, he got a chance to know his neighbours and build a history together' (Cheney, 2007: A1). While suburban spaces may not be physically accessible because of the sheer sprawling spatial extent of the suburbs, they have the potential to be socially and financially accessible to a diverse range of people. The suburbs are the places where 'spontaneous gesture[s]' (Lerman, 2002) of vernacular creativity can, with perseverance, be uncovered.

The inner suburb of Etobicoke: a case study of vernacular suburban creativity

Canadian cultural workers are predominantly concentrated in urban locations (Bunting and Mitchell, 2001). A report commissioned by the Canada Council for the Arts, entitled *Artists in Large Canadian Cities* (Hill Strategies Research Inc., 2006), provided a statistical portrait of arts employment in 92 Canadian cities with populations over 50,000. The report determined that in 2001 there were 130,700 artists in Canada, earning on average $23,500 (26 per cent below overall labour force earnings), and representing 0.8 per cent of the labour force. Between 1991 and 2001 the number of artists in Canada grew 29 per cent, three times the rate of growth in the overall labour force, with the majority of these artists living in one of Toronto, Vancouver, or Montreal. Toronto is Canada's largest-census metropolitan area, and in absolute numbers it has the largest concentration of artists. According to Statistics Canada data, the number of arts and culture professionals in the Toronto city region grew from 17,200 in 1996 to 48,575 in 2006, an increase of 182 per cent, with over 13,000 living outside the City of Toronto in surrounding suburban and exurban communities. The inner suburb of Etobicoke, on the western edge of downtown Toronto, had 2,280 individuals employed in professional occupations in art and culture in 2006, up 15 per cent from 1,985 cultural workers in 1996. The interpretation of 'inner' and 'outer' suburbs employed in this chapter is informed by the work of Bunting *et al.* (2004), who use the age of housing stock to make the distinction. In the Canadian context, an 'inner' suburb has housing stock built between 1946 and 1971 and an 'outer' suburb has housing stock built after 1971.

In 1996, The City of Toronto Act created a new municipality by amalgamating the Borough of East York and the cities of Etobicoke, North York, Scarborough, York, and Toronto into a megacity. Amalgamation made the City of Etobicoke part of the City of Toronto. However, few Etobicoke residents describe themselves as living in Toronto – they remain loyal to Etobicoke as a place identifier. But Etobicoke is far from being a unified community. Instead, Etobicoke is fractured into three districts that span the socioeconomic class spectrum from extreme wealth to poverty, with very little in the middle: North Etobicoke, Central Etobicoke, and South Etobicoke.

North Etobicoke (Rexdale) is a neighbourhood of high-rises, subsidized housing, and a large immigrant population. The media have strongly associated this neighbourhood with the negative forces of urban decay, gun violence, drugs, and gang activity. It is one of Toronto's 13 'priority neighbourhoods' identified in Mayor David Miller's community plan and has been targeted for revitalization through projects that combine culture and recreation. As the following two quotations suggest, few arts practitioners in Central and South Etobicoke have visited North Etobicoke or are aware of what cultural resources and facilities are available there: 'I don't even know where North Etobicoke is. I mean I think it's that bit way up there kind of north of Dundas and Bloor maybe?' (Faculty member, Humber College, 16 November 2006). And 'I haven't gone and visited North Etobicoke because it's further out. I don't know what's there' (Visual artist, Lakeshore Artists Village Cooperative, 17 November 2006). In the social imaginary, Etobicoke is a sprawling and fragmented place with no clearly defined centre.

Central Etobicoke (Kingsway) is a predominantly white, upper middle-class neighbourhood centred on Bloor Street West and the Kingsway. Because it has a wealthy and highly educated population, a significant amount of arts consumption takes place here. As one long-term resident described Central Etobicoke: 'This area is not an area of colour. You do not find many Black people or Asians. That's comparatively rare. But you know, the further north you go and the further south you go, the less that is true ... the Kingsway is a very expensive neighbourhood and it tends to be people who have been here a long time. People just take their parents' homes and live in them or sell them off. But I think it's a lot of old money' (Composer, Etobicoke, 9 November 2006). Central Etobicoke is an island of leisured wealth among a sea of working-class poverty.

South Etobicoke (The Shore) can be described as a geographically isolated sliver of an island. It is framed by Etobicoke Creek and Humber River on the east and west, Lake Ontario on the south, and a river of concrete and rail, the Gardiner Expressway, the Go Train line, and large or abandoned factories, on the north. As one arts administrator and long-term resident explained, 'Like islands, Lakeshore has a split personality. On the one hand it's fairly insular. You know, there's a sense of you know people who come from away are not to be trusted. But because we're an island, we're fairly self-sustaining and we're fairly close. So, there's a strong sense of you couldn't pull any stunts in this neighbourhood because somebody would know and they'd tell somebody else' (Faculty member, Humber College, 4 October 2006). Part of the small-town feel comes from the fact that South Etobicoke is made up of three neighbourhoods with long histories as towns in their own right. From east to west, Mimico, New Toronto, and Long Branch, retained their town status until 1967. In the 1890s, Mimico was where some of Toronto's wealthy industrialists built their summer homes, New Toronto was a waterfront community built around the railway line with heavy industry and housing for working-class factory workers, and Long Branch was a summer resort town of cottages, hotels, amusement rides, and a boardwalk. In the present day, South Etobicoke has become incredibly ethnically diverse, with high numbers of Korean, Spanish, Polish, Lebanese, Caribbean, and East African residents. It remains predominantly working class. There are pockets of middle-class wealth in older homes along the shoreline and in new condominium and townhouse infill. Much of this residential intensification has been constructed over the last five years as part of the city's smart growth policies designed to combat urban sprawl by diverting new Greater Toronto Area residents into the inner suburbs. This new residential construction has gradually brought in people with a greater disposable income and has begun to change the retail mix – the Dollarama, the pawn shop, and greengrocers have been joined by an organic butcher, a food caterer, and high-end coffee shops (that exhibit art) and restaurants. South Etobicoke has more artists and arts activities per capita at a community level than does either Central or North Etobicoke. There are several reasons for this, which will become apparent in my ensuing discussion of some of the advantages of engaging in a suburban creative practice.

The advantages of the suburbs

The decision to undertake cultural work in the suburbs can be influenced by a range of different factors. In Etobicoke, four variables became particularly apparent in interviews with arts administrators for commercial and not-for-profit arts organizations, executive members of arts councils, cultural planners, and arts practitioners (e.g. visual artists, dancers, writers, actors, musicians, conductors etc.): natural setting; housing affordability; downtown (dis)connection; and cultural 'popcorn' pioneering in multi-purpose spaces. Each of these will be briefly discussed in turn.

Natural setting

The idea that natural environments foster cultural and intellectual creativity has a long history within the arts. Seasonal or periodic relocation to a more natural setting for inspiration – an artists' retreat – is certainly a common phenomenon in the visual, musical, and literary arts. Many of the cultural workers interviewed expressed a romanticized understanding of the inner suburb as a place that was close to nature and away from the tension, intensity, and frenetic pace of urban life. They repeatedly highlighted the importance of living and working in close proximity to Lake Ontario, to parkland and to open space because it allowed them to commune with nature in relative quiet and isolation. As one photographer explained:

> Those areas below the Lakeshore, I find them calming. The lake is down there. I bicycle a lot. So, I sort of like the idea of being able to get out of that urban core to just relax and decompress a little … The biggest thing that I like around here is the lake and Colonel Sam Smith Park. I love the fact that there's hardly anybody in there and on some level you can go there on

a Sunday afternoon or morning and you're just flying around and you think you're a long ways from a city core. The Etobicoke Creek, you go in there and you feel like you're 200 km away from Toronto.

(Photographer, 8 November 2006)

It is apparent in this quotation that the suburb is understood by this cultural worker in quite a traditional sense as a place that dispels tensions and produces spaces and times of calm in the midst of the chaos of urban living (Pile, 1999). Other arts practitioners go on to identify the parkland landscapes as sources of inspiration. One printmaker has spent the last several years working on a botanical series inspired by the plants she sees when cycling and walking through the local parks: 'I'm two blocks from the lake and I love walking down there. When I can, I ride my bike ... the lake, the water, the clouds and the plant life influence my art work, which I don't think I would have been inspired by if I wasn't so close to it' (Visual artist, 25 October 2006). The water is a particularly influential feature of the natural landscape that is present in the paintings, prints, plays, and music produced by local artists, many of whom frequently walk and bicycle through the parklands along the shore of Lake Ontario.

The freedom to explore, without permission or financial expense, appeals to cultural workers. They can pretend that this parkland is a wild natural landscape far away from the city and the forces of consumption and production, rather than a place where nature is manicured and artificial. That sense of inspiration and escape is neatly expressed in the comments of this musician:

The peace and quiet of the area expresses itself in my music ... what it does is it cleanses your brain and allows the muse to happen. There's just something very therapeutic here. If this were the morning I'd say 'You want to go into the midst of nowhere, I can get you there in a 10 minute walk'. It's incredible the unexplored territory down here.

(Musician, 24 October 2006)

Proximity to nature and separation from city life are themes that characterize local artists' appreciation of suburban life in Etobicoke, but they stand in marked contrast to the industrial, working-class heritage of the suburb and the vast distance that separates the central and northern portions of the suburb from the lake. It is curious that artists are inspired by the natural environment but not by the working industrial environment. In the central city, living and working in former industrial districts in reclaimed factory warehouses is glorified and romanticized (Bain, 2003; 2006). However, in the inner suburbs, industrial landscapes are characterized as bleak, dry, flat, tired, and polluted; they are far from being sources of inspiration or places of romanticization. Cultural workers can look south to appreciate nature, but do not look at, travel to, or explore the industrial lands and transportation corridors to the north, the manufacturing heartland of both Etobicoke and the city of Toronto. The relics of industry seem more interesting than the harsh lived realities.

Housing affordability

A close second to the natural setting when cultural workers discussed the appeal of Etobicoke was affordable housing within close proximity to downtown Toronto. Bunting *et al.* (2004) have identified that the unaffordability of housing is a growing problem within the fastest growing Canadian Census Metropolitan Areas because these have inflated and exclusionary housing markets. High rent and purchasing costs within Toronto's central city have pushed cultural workers out into the inner suburbs, where real estate is more affordable. A house near the lake in the established and gentrified east Toronto neighbourhood of The Beach would likely cost upwards of $750,000, but in South Etobicoke it can be had for closer to $400,000.

The affordability of real estate is a strong pull-factor for professional artists and has played an important role in creating a concentration of cultural workers. As one visual artist explains: 'South Etobicoke has musicians, writers, film people, actors. It just seems to be a very creative kind of place. It's very much like the Beaches, there's a lot of creative people living here. I think basically because you can get low cost housing but you're close to downtown. You're 20 minutes by car, you're 40 minutes by streetcar or subway' (Visual artist, 5 November 2006). It becomes relatively easy, then, for professional visual artists, musicians, and actors to live in Etobicoke but to not be forced to struggle to make a living there. Instead, they can commute into downtown Toronto by public transit for their bread-and-butter jobs or necessary cultural amenities and resources.

Downtown (dis)connect

Downtown Toronto is the locus of the majority of facilities and organizations that provide services to cultural workers in the Greater Toronto Area. Many Etobicoke visual artists, actors, and musicians retain memberships with downtown arts organizations, which enable them to receive weekly newsletters and electronic updates on workshops, performances, and exhibitions. Thus Toronto functions as an information hub – knowledge is collected, made significant, and then transferred out to the suburbs. This distribution function helps to reinforce Toronto as the creative focal point of the city region. Etobicoke's proximity to downtown Toronto, then, is a double-edged sword; while the central city provides valuable cultural sustenance, it can also draw away cultural vitality from the inner suburbs and foster a relationship of cultural dependence. This relationship reinforces the mistaken assumption that if cultural activities occur outside of the downtown core, then they are not as significant or of as high quality.

Etobicoke, however, began life as a city, not as a suburb. Consequently, the mindset of being on the edge of where cultural activity happens is divided. As one arts administrator explains: 'I mean part of our mind knows that we can hop in the car and go downtown to visit the national this, the Canadian that, and have some amazing experiences and see some amazing work … On the one hand, yes, we sometimes think that we're in the 'burbs and we have to go downtown for our cultural fodder. But by the same token because Etobicoke was its own city it had cultural activities happening here' (Faculty member, Humber College, 4 October 2007). But those cultural activities get very little recognition from central city arts organizations, attract few audience members from downtown Toronto, and receive minimal local and city-wide media coverage.

The perception of cultural and physical distance between Etobicoke and downtown Toronto is distorted. On the one hand, local cultural workers regard the central city as a close commute and as a valuable source of cultural resources. On the other hand, they realize that many central city residents perceive Etobicoke in a very different light, as a place that is far removed. As one artist explained, when 'you talk to someone about where you have your studio they'll say 'Way out there?' and I'll say 'Well, the Queen streetcar goes right by the end of my street. So, yes, there is that perception of it being sort of far away' (Visual artist, 25 October 2006). In conversation, Etobicoke is frequently casually referred to as 'out there' on the urban periphery, in a cultural wasteland. 'People in the city see the west end, see Etobicoke, as no man's land. If you say to someone at Yonge and Bloor, "Let's go to Lakeshore and Kipling" they'll say "Are you insane? Why would I go there? Is there anything there?"' (Director, Lakeshore Arts, 27 September 2006). Etobicoke is on the geographical margin at the edge of Toronto. Some cultural workers dismiss this edginess as being 'annoyingly cut off' and fostering feelings of insignificance, while others embrace it for the sense it affords of being away from the pressures and competition of the mainstream, and the creative possibilities that distancing affords.

Cultural 'popcorn' pioneering in multi-purpose spaces

Etobicoke is located off the cultural radar, in an untrendy, 'not-yet-officially-designated-creative' part of the city. In interviews, some cultural workers expressed an appreciation of the sense of escape it afforded them, the opportunity to get their hands dirty and to make something from nothing. A photographer who relocated from New York City and then again from downtown Toronto wants to capitalize on the untapped potential he sees in Etobicoke:

> I still subscribe to that theory that if you build something unique, some people will come. They'll come locally and you could even draw some people from another place. Now, maybe even downtown where everything is basically established. There's so many people that have already manifested these kind of ideas downtown, there's not much room for more. Whereas out here, this is sort of like the Wild West, where you can start something.
>
> (Photographer, 8 November 2006)

This reference to the frontier motif is one that I first uncovered in my doctoral research, interviewing contemporary Toronto visual artists who lived and worked in inner-city, working-class neighbourhoods that had minimal exposure to the arts (Bain, 2003). Only now, the cultural frontier has moved further out. But the experience of artistic isolation, financial hardship, and potential for neighbourhood involvement remains the same. The pioneering qualities that are being celebrated also remain similar: individualism, independence, self-sufficiency, thrift, and courage to take risks.

The cultural pioneering that is taking place in Etobicoke is happening largely out of necessity. Simply stated, Etobicoke lacks cultural focal points. There are few purpose-built venues in which to partake of cultural activity. Those venues that do exist are, for the most part, multi-purpose buildings: high school auditoriums (e.g. Etobicoke School of the Arts – but only a small percentage of the students are from Etobicoke), churches (e.g. St. Margaret's Church; Mimico Presbyterian Church), community centres (e.g. Franklin Horner Community Centre), libraries, shopping malls, and the civic centre. Churches in Etobicoke are often used as performance venues, particularly for theatre and music. They can be rented inexpensively on an hourly basis with no financial overhead. Church boards are often amenable to this sort of creative use of their facility because they have a mandate to reach out into the community and to bring residents together. One of the drawbacks, however, is lack of storage space – additional storage facilities are often rented by artists and arts organizations off site. Libraries are also widely used to run arts work-shops and community events. With few, if any, commercial galleries, a handful of coffee shops have become some of the only venues locally to exhibit fine art (e.g. Rocket Fuel Shop; Roast Beanery). As the Director of Arts Etobicoke (15 September 2006) explains: 'There are no museums. There are no large galleries. There are no performing arts centres. There's no concert hall. You know, there's nowhere to go.' The key cultural facilities in Etobicoke, with the exception of the Neilson Park Creative Centre, are mostly concentrated in the south, along the Lakeshore (e.g. Humber College, the Assembly Hall, and the Lakeshore Artists Village Cooperative), and are designed to support a range of different activities.

Take, for example, the Assembly Hall. Located south of Lakeshore, beside Humber College, it is a city-run heritage property opened in 2001 that cost $4.5 million to renovate, and is the only community-based cultural centre in the west end of the city. The mandate of this facility is to function as 'a gathering place at the heart of the Lakeshore where community and creativity are celebrated and nurtured' (Manager, Assembly Hall, 13 November 2006). When asked how 'crea-tivity' was understood in this mission statement, the manager went on to state: 'Creativity is anything that's somehow related to arts … So, anything where either people are coming to see other people's creative efforts or are coming to do some-thing that is creative themselves.' This is a rather vague interpretation of crea-tivity, but one that perhaps accurately reflects the range of uses supported by this building. Intended mostly for theatre, the performance hall in Assembly Hall seats 250 people and is rented out on a sliding fee scale to community theatre groups, to Humber College for student performances, and to individuals for private events. While there is a stage, there are only two small dressing rooms. The acoustics in the performance hall are not ideal for music. Fine art is displayed in the left-over spaces of hallways and foyers, spaces that get very little people traffic unless a performance or meeting is scheduled. With over 300 permits issued a year, the facility hosts everything from cake decorating to children's drama classes and it is not used strictly for arts programming. 'Because it is one historic building that was carved up, almost everything is multi-purpose, multi-usage. It's a wedding today, it's a performance tomorrow, it's a community meeting the next day. There's not full amenities for anyone' (Manager, Assembly Hall, 13 November 2006). This blending of activities, the manager believes, will bring culture to a wider audience who might otherwise not cross the entrance threshold.

Another example of a multi-purpose space that supports more than the arts is the Franklin Horner Community Centre. This is a unique, alternative incubator space. A hundred-year-old former elementary school building that was converted into a community centre in the 1980s, it has gradually evolved into a social, recreation, leisure, and sports facility for people ages 0 to 100. Its membership base of 1,150 is drawn from the Filipino, Polish, Italian, Serbian, Mexican, and Anglo communities. Without permanent city funding, this community centre relies on many of these members, particularly those who are older adults, as volunteers to renovate and repair the building and to help run fundraising events. Fifty-three member groups are based here, including: Lakesteps Narcotics Anonymous, Ontario Early Years daycare program, Colin's taxi driving school, the Humber Barber Shop Quartet, the Polish Eagles dance troupe, and the Filipino Folklore society. Indian wedding and anniversary parties are held in the gym, along with after-school tutoring and sports programs; Mexican dance troupes train in a dance studio with specialized sprung floors that are also shared by yoga students; a 17-piece band practices in the auditorium; seniors' lunch-and-learn programs happen in the craft room; the woodworking machine shop is busy all week; a former girls' bathroom has been converted into a darkroom for photography classes (and featured in a British fine art photography magazine); fine arts studios are rented on a permanent basis by professional artists as well as used on a more temporary basis for photo shoots and life-drawing classes; and the local arts service organization, Lakeshore Arts, has moved in as a cost-saving initiative. The long-term goal of the manager of the Franklin Horner Community Centre is to make this an arts teaching facility, with a name such as the Franklin Horner Fine Arts School. It is these sorts of initiatives that leave suburban cultural workers optimistically saying: 'it's like popcorn, I hope. A couple of things pop and then all of a sudden pop, pop, pop. Will it ever just boom, I don't know. But I'm starting to notice popping' (Photographer, Etobicoke, 8 November 2006). Etobicoke has not yet reached what Malcolm Gladwell (2000) has referred to as 'the tipping point', but change is certainly underway in the suburban landscape.

Conclusions: appreciating the variegated spatialities of creativity

Where creativity happens matters to how it is expressed and to how it is understood. Cities and city centres have long been privileged as the singular great locations of technological and cultural innovation, progress, and creativity (Nicolaides, 2006; Perl, 2001), and celebrated as the places where 'risks are taken, problems are raised, experiments tested, ideas generated' (Urban Cultures Ltd, 1994, cited in Montgomery, 2008: 31). Urban creativity is usually associated with the trendy 'edginess', spectacle and risk taking of an evening economy that caters to the creative class: 'a mix of old and new buildings, an active streetscape, mixed use, contemporary design, cafés and bars, [and] nightclubs' (Montgomery, 2008: xxv). Without the density, fine grain, and cachet of the central city, the suburbs with their piecemeal sprawl and segregated land uses are often too readily dismissed as uncreative. Yet, as this chapter has revealed, creativity comes in many different forms and is stimulated by many different environments – not all of which are large urban centres. The suburbs of Canadian cities are also uniquely textured with their own intensities, elasticities, and complexities.

The Etobicoke case study has illustrated some of the ways in which suburban creativity is different from the urban creativity that is usually concentrated in the central city art districts of large metropolitan areas. Suburban creativity operates and mobilizes on a different spatial scale with substantially less scholarly, political, financial, and public support and recognition. Moreover, suburban creativity can be a more interdisciplinary, social, collaborative, inclusive, flexible, and communal process that is often driven by grassroots organization and is frequently located in multi-purpose spaces. Taken together, the freedom to explore green space without permission or financial expense, the affordability of housing, the spatial (dis)connect from the established networks and expectations of downtown arts scenes, and the lack of purpose-built cultural venues have fostered creative opportunities for cultural workers in the suburbs. In place of completeness and closure, suburban cultural workers value the unpredictability of new uses and possibilities sheltered in the improvisational spaces of the suburbs.

Suburban improvisational space supports small-scale experiential initiatives and local interventions in the urban fabric that 'not only … disrupt the authoritative structures that govern [space], but also … encourage a dialogue about the possibility for other forms of being and behaving' (Jonsson, 2006: 37). As an in-between space that holds the potential and possibility of plurality and contradiction inherent within it, suburban improvisational space could be interpreted as a more inclusive alternative to the spectacular spaces of urban creativity (Bain, forthcoming). Such localized 'spaces of hope' (Blomley, 2007: 59) or 'Temporary Autonomous Zones' (Routledge, 1997) are the sites from which grassroots activism, optimism, and creativity can emerge that resist dominant ideologies and practices.

If creativity is to remain a viable and inclusive tool of urban economic development, then scholars, policy makers, journalists, and the general public need to renew their stock of collective imaginings of creative practice. Such re-imaginings of creativity need to include the suburbs and the seemingly unspectacular and local community arts interventions that do not always garner media attention but have the potential to engage socially, culturally, and economically diverse audiences. The spaces and expressions of suburban creativity need to be valorized, and less readily dismissed as uninteresting and technically and conceptually naïve. In an era of neo-liberal, expert-driven models of cultural policy development that fetishize downtown artistic enclaves as the epitome of urban coolness and celebrate middle-class forms of cultural capital, garnering respect, attention, and financial investment for suburban creative practice is likely to be an uphill battle – but it is a battle worth fighting.

6 Beyond bohemia

Geographies of everyday creativity for musicians in Toronto

Brian J. Hracs

Bohemia is a lifestyle with thematic elements that can be read through very practical instantiations in a range of urban contexts since its Parisian origins. Both the idea of bohemia and its associated spatial practices have proven durable and portable, which is evident in cities throughout Europe and the United States. But as we will see, each bohemian eruption is both familiar and quite distinctive because of the material and spatial specificities that it encounters in a particular city at a particular time (Lloyd, 2006: 54).

Since it was first used to describe the lifestyle of eccentric artists in the 1830s, the notion of bohemia has served to connote alternative living. Today studies suggest that the geography of bohemia is highly concentrated in large cities (Florida, 2002). Even as new bohemian neighborhoods unfold in a dynamic urban landscape, significant continuity is said to exist between these communities and their counterparts of the past (Lloyd, 2006: 69). Indeed, bohemian spaces continue to be characterized as cheap, gritty, dangerous and isolated, and these features help to attract traditional bohemians, including artists and musicians. As new technologies, techniques and communication networks facilitate creative practice in a growing range of sites, however, these highly concentrated pockets of creativity are spilling over from downtowns into vernacular spaces in suburbs. In particular, there is evidence that the changing nature of independent music production is becoming increasingly difficult to reconcile with the romanticized milieu of bohemia.

In this chapter, I argue that although creative activity largely remains clustered in the downtowns of cities and many artists still choose to pursue bohemian lifestyles, the employment conditions associated with independent music production have caused some musicians to reject bohemian spaces. More specifically, I demonstrate that, in order to achieve the most favorable balance between the cost, location and characteristics of their life/work spaces, some musicians in Toronto are relocating from the bohemian enclaves in the downtown to 'everyday' locations in the inner and outer suburbs. In Toronto rings of automobile-dependent suburbs surround the downtown core. The inner suburbs were constructed between 1946 and 1980 and the outer suburbs have been constructed since 1980 (Noble, 2008). Located immediately outside the downtown, the inner suburbs have lower rents than do the outer suburbs, although these, too, usually remain more affordable than the downtown. In particular, these vernacular spaces of creativity include ordinary, functional and 'square' houses, basement apartments, converted garages, churches, retail spaces and small recording studios. In addition, I argue that a growing range of push and pull factors, such as the cost of living, competition for employment and local 'buzz' influence the spatial patterns of musicians. Moreover, I will show that some of the key features of bohemian living hinder the creative process. As a consequence, I suggest that by privileging downtown clusters as the only sites of creativity, existing academic studies and policy initiatives fail to recognize the increasingly important creative outputs emerging from everyday spaces outside of the core.

The findings presented in this chapter are drawn from 65 semi-structured interviews conducted with musicians and key informants in the music industry in Toronto, including executives at major and indie record labels, studio owners, managers, music professionals and government officials. To reflect diversity, the respondents include participants in Toronto's jazz, rock, punk, electronic, hip hop and classical scenes. Beyond genre and employment status, the musicians interviewed are also differentiated by stage of career, place of birth, age, gender and level of education.

The chapter begins by reviewing the traditional connection between artists and bohemian spaces. This is followed by a section which outlines how recent changes to the music industry are affecting the working lives and creative preferences of individual musicians. The next section explores some of the factors that serve to 'push' musicians out of Toronto's bohemian enclaves. The final, empirical section provides an analysis of the 'pull' factors currently attracting these displaced musicians to the suburbs. I conclude by considering the implications of this outward flow of creativity.

Artists and the allure of bohemian neighborhoods

The concept of bohemia emerged in the 1830s to describe the activities and lifestyles of artists and other eccentrics in the Parisian arcades. Over time, subsequent bohemian communities have formed in North America in sites such as Greenwich Village and Soho in New York, Wicker Park in Chicago and Queen West in Toronto. Such bohemian spaces are often characterized as densely populated, rundown and dangerous and are often located in the most undesirable and isolated quarters of the city. By extension, it is the affordability, grittiness and isolated nature of bohemian neighborhoods which attract artists, including musicians as well as visual artists, writers and dancers. Accounts of bohemia, for example, describe artists as embracing the creative stimulus associated with derelict and dangerous spaces and romanticize the notion of 'grit as glamour' (Lloyd, 2006). For the artists in Bain's study of Toronto, the violent backdrop of drunken street fights, homicides and prostitution represents a boundary that isolates artists from mainstream society (2003). More broadly, artists who live in these bohemian neighborhoods are often characterized as fearless urban pioneers whose courage, tenacity and practicality are celebrated as they carve out live/ work spaces in these danger-filled battlegrounds. For instance, in his case study of New York's Lower East Side, Smith also makes use of frontier imagery, describing bohemian spaces as a glamorized landscape of frontier danger and savage energy (1996: 18).

Accounts of bohemia also explain that artists need to isolate themselves from mainstream society because their temperaments and lifestyles, which include a desire for nonconventional sexual norms as well as liberal use of drugs and alcohol, are antithetical to bourgeois conventions. In particular, Lloyd notes that 'the rationalized organization of labour and commerce was anathema to Bohemian sensibilities' (2006: 60). The creative process, therefore, is portrayed as dependent on the stimulation from bohemian spaces and fundamentally incompatible with the banal, standard and 'square' aesthetics found in everyday suburban spaces.

Within these bohemian communities, artists are said to make extensive use of 'third spaces' such as coffee shops to develop creative ideas and network with other artists. As Lloyd notes, the Urbus Orbis coffee shop in Wicker Park, Chicago enjoyed immediate patronage from the nascent arts community in the early 1990s and provided a site for artists to just 'hang out' while waiting for the lightning bolts of inspiration to strike (2006: 108). As these accounts emphasize the attraction to gritty spaces and the sites of networking, we are left with the perception that artists in bohemian communities hang out in 'third spaces', sipping bottomless cups of coffee and going about the creative process in a relaxed state. As described, therefore, the bohemian lifestyle is seemingly devoid of the structure, professionalism and time constraints experienced by other members of the labour market. However, the following section demonstrates that, for some musicians, changing employment conditions are weakening the ties to the spaces and ideals of bohemia.

The changing nature of employment for musicians

Musicians are at the forefront of recent changes to the way that cultural products are created, distributed and consumed. At the macro-scale, the music industry has been thrust into the digital age and forced to deal with the growing specter of internet piracy and the protection of copyrights and intellectual property. At the micro-scale, the employment structures for individual musicians have been radically altered by the rise of independent production. While many of the causes and consequences of this digital shift have been addressed in the literature (Leyshon, 2003; Leyshon *et al.*, 2005; Power and Hallencreutz, 2005), the impacts on the working lives of individual musicians are less well understood. In particular, there is a need to consider how new employment structures have altered the employment trajectories and residential preferences of musicians and the extent to which bohemia is still a dominant feature of their lives.

In the wake of a technologically induced economic downturn, the major record labels have terminated many musicians' contracts and have reduced the services and resources available to those who remain signed. One respondent, who works for Universal Music Canada, for example, reported that the number of new contracts handed out by the major labels in Canada has been reduced to four new contracts per year. While the majority of the revenue generated from music is still attributed to the major labels, a majority of musicians now operate independently of these major firms. According to the Canadian Independent Recording Artist Association (CIRAA), fewer than 5 percent of the musicians in Canada are signed to recording contracts, making the remaining 95 percent of musicians, by definition, independent. As one musician explains:

> In the early 1980s, being an independent musician was a choice, some people didn't want to work towards a major label deal because there were restrictions and conditions attached to that ... Now very few artists can still get signed to major label deals, so the majority of artists end up on the independent side.
>
> (Indie musician and music producer, July 2008)

Consequently, independent or 'indie' music production has been transformed from a niche alternative to the dominant structure of employment. Independent musicians are now responsible for all of the creative, technical, managerial and business tasks individually and this has fundamentally altered the way musicians approach their careers. Under the 'indie' model, respondents reported the growing need to become more efficient and professional. Accordingly, creative tasks such as song writing and rehearsing constitute a shrinking fraction of a musician's day. Instead of lounging in coffee shops, independent musicians in Toronto work long hours performing non-creative tasks, as this musician and educator explained:

> If you actually want to make a living as an indie musician, it is a tough go. You've got to pretty much do it yourself all the way through. You have to be able to play your instruments well, write songs, but you also have to be able to get out of the basement and perform them ... You also have to be a booking agent ... you have to be a manager, setting up interviews and getting the word out ... You also have to raise money and get financing together to do some recording, so that means grant applications, going to the bank and putting together business plans and proposals ... Plus there are all the technical skills that you need. How to put together a home studio, how to get good recordings. What is involved with recording and mixing and mastering ... If you are going to put out an actual CD then you need to have some kind of artwork with that as well. Marketing is another one, getting lists of media that you can approach, radio stations and magazines, fanzines that you can send your music to for review, all that kind of stuff and promotion. Merchandizing, maybe it is just going to be T-shirts, but often it is much more than that now, and these are all things that would be done for you by various people in big organizations if you were signed to a label, but now you have to do all of these things yourself ... So musicians are now responsible for the whole range of activities, technical, business, performance, musicianship, you have to have it all together.
>
> (Indie musician and educator, July 2008)

Beyond requiring musicians to combine their creative talents with new technical, managerial and entrepreneurial skills, independent production in the digital age also requires high degrees of professionalism, efficiency and organization. However, as these traits and behaviors run counter to the alternative ethos of bohemia, in order to survive some musicians are being forced to adopt the square, self-reliant, standardized and sterile lifestyles closely associated with Whyte's 'organization man' (1956). In particular, the example of networking illustrates this shift, as encounters between musicians have become less organic and now tend to be more structured, with specific schedules and agendas. More specifically, my research indicates that the relationships between musicians and other 'creatives', including fashion designers, photographers and web designers, have moved beyond the barter exchanges commonly found between artists in bohemia and now resemble business transactions between firms. With hectic schedules, 24/7 connectivity and BlackBerries in hand, 'indie' musicians are losing the luxury of bohemia, no longer able to reconcile the demands of independent music production with hanging out in a romanticized milieu.

In addition to these structural changes, technology has also afforded individual musicians unprecedented geographic mobility. No longer tied to the major labels and established sites of music production such as Los Angeles, New York and Nashville, independent musicians are now free to live and work wherever they choose. While the majority of musicians still choose to locate in major cities, technology allows music production to take place almost anywhere. Being in a central location within the city is no longer essential. As one musician put it:

> Proximity is not paramount, you can be in the Arctic or anywhere with a wireless connection and conduct your business.
>
> (Multi-instrument musician, May 2007)

With independent production, the working lives of musicians have become more professionalized and distant from those of traditional bohemian artists. Consequently, a growing number of musicians now resemble business-minded entrepreneurs who make calculated decisions about their careers, where they live and how they use neighborhood spaces. Instead of articulating a common penchant for bohemian living, the musicians interviewed in Toronto based their career and life decisions on different criteria, including demographics, experience, preference, values and life-cycle attributes. As one musician explained:

> Years ago there was more homogeneity in how you defined life as a musician. But these days each of us puts things together in such a particular way that our checklists for living and working are very different.
>
> (Guitarist, September 2008)

The decline of bohemia

My interviews with musicians confirm that vibrant bohemian quarters still exist in the downtown core of Toronto and that many musicians and artists still prefer to live in these neighborhoods. Increasingly, however, the decision to locate in the core is being made for lifestyle considerations rather than employment or creative requirements. As one musician put it, it is no longer crucial to live in the down-town to be a musician:

> I think living in the city is a personal choice. It is not for music, although seeing shows is good. So there is a distinction between lifestyle and what you actually need to be effective as a musician. So living in the suburbs would not hinder anything musically, but I would be bored.
>
> (Singer, May 2007)

In particular, two groups of individuals prefer to live downtown. The first include very young musicians, often from smaller urban settings, who are trying to plug into networks, learn the ropes and make a name for themselves. For these individuals sacrificing space in favor of proximity is regarded as a necessary trade-off to succeed in their musical careers. This musician, for example, spoke of the importance of connecting to the downtown scenes:

> At the beginning stage of a band I think it is really important. Until you meet people and have people working for you it is really important to be seen and make your mark and solidify yourself.
>
> (Guitarist and singer, September 2007)

The second group come from the other end of the spectrum, namely, older more established musicians who have learned to successfully manage the risks of inde-pendent music production. For these individuals the preference for downtown living, however, is not predicated on the local buzz or nightlife, but rather on proximity to their customary sites of work and collaborators. Moreover, many of these musicians secured affordable and artist-friendly space before Toronto's real-estate prices skyrocketed. Despite living in Kensington Market, one of Toronto's most identifiable bohemian enclaves in the city, this musician based his choice of location on proximity rather than on any connection to bohemia itself:

> I'm an older guy and a privileged guy who owns a house and I have a partner who's got a steady job and between the two of us we were able to buy a house ... It's huge. I need that. I need a place to store my equipment, to be able to play, to be able to record, and then be able to get my gigs easily ... So this location works for me for my work. My steady gigs are downtown, most of my club gigs and most of my corporate gigs are downtown. When I teach, I teach in my house. Most of the recording I do is either at my house or at somebody else's studio, which is usually close to here. For the work that I do it's much easier for me to do it here.
>
> (Guitarist, September 2008)

In between these groups a growing number of musicians reported being disen-chanted with bohemian living. For these musicians, the allure of inhabiting decaying urban frontiers had worn off and the grit, danger and isolation of bohe-mian spaces were cited as 'push' factors. Crucially, these aesthetics were also described as counterproductive to the creative process and career paths of musicians. As the following sections will demonstrate, the changing nature of employment and increased mobility afforded by technology has caused many musicians in Toronto to rethink their residential and work choices. More specifically, rising rents, overcrowding, competition and the negative externalities associated with local buzz are pushing some musicians out of Toronto's downtown core and into the city's inner and outer suburbs.

The pursuit of practicality

The artistic requirements of affordable, flexible and centrally located live/work spaces provided the original impetus for the conversion of derelict industrial 'loft' spaces (Zukin, 1982). Indeed, interviews with musicians who originally migrated to Toronto in the mid-1990s confirm the attraction to affordable space in bohemian quarters and suggest that, for those willing to sacrifice, such space was available within the downtown core.

> The reason why I moved [to this house] was because I got a room ... like a pantry, it had no heat or a window ... my rent was $150 [a month] for a long time. But yeah, if I had to pay $800 rent I would have been on the street in two seconds or have had to move back home with my parents in Guelph.
>
> (Multi-instrument musician, May 2007)

As the city's real-estate market has taken off, however, the artistic enclaves located in the downtown core have been increasingly threatened. Consequently some musicians have been priced out of the market. As this musician explains, gentrification in her neighborhood forced her to relocate from the downtown to the inner suburbs:

> I lived in a very roomy apartment for 24 years and it was so spacious, and had a great view, and really cheap rent, but we finally had to leave that and move into this tiny house because the owners decided that they wanted to renovate and raise the price.
>
> (Pianist, July 2008)

Despite his desire for a central location, this musician explained that, as prices continue to rise, finding suitable and affordable space is very difficult:

> I have always wanted to live as close to the action as possible but in the mid-90s it just became too expensive to live downtown so I moved out to the west end, which back then was the wild wild west ... I realize that living centrally is kind of a luxury.
>
> (Musician and music programmer, November 2007)

Beyond proximity to performing venues, however, few participants spoke positively about other quintessential features of bohemia and several musicians commented that they preferred safe and clean spaces in the inner and outer suburbs to the gritty and dangerous bohemian enclaves:

> You can live in suburbia or the outer parts of the city, because downtown Toronto is pretty dirty, there are too many bums and crack-heads. The rent is ridiculous, and it is quieter outside of the city.
>
> (Guitarist, May 2007)

As another musician put it:

> I would rather live in a safer or nicer area than right where the scene is if it's really run down or dangerous.
>
> (Drummer, March 2007)

Earlier, I argued that the demands of independent production have forced some musicians to adopt professional traits and practices associated with mainstream corporate culture. The musicians rejected the alternative ethos in favor of 'square' business practices, in order to survive. The quotes above further suggest that some musicians prefer 'square' aesthetics and lifestyles to those of bohemia. In particular, the quieter, safer and cleaner spaces found in the 'everyday' suburbs are more attractive than bohemian spaces and lifestyles.

The limits of local buzz

Some musicians are also rejecting the benefits of co-locating in densely populated creative communities. Counter to the logic found in the economic geography literature on the importance of 'local buzz' (Bathelt and Malmberg, 2004) and 'being there' (Gertler, 1995) the overpopulation of musicians in Toronto's downtown core has produced a series of negative consequences. My research suggests that Toronto's live music scenes have surpassed a sustainable threshold and the consumer market is no longer large enough to support the number of musicians working in the city. Competition for the few well-paying employment opportunities is intensifying and many musicians, who already earn low incomes, are being forced either to play for free or, worse, to pay the venue owners to get on stage. (The 2001 Canadian census reports that musicians earned average annual incomes of $16,090, or 75 per cent less than the national average.) These conditions are exacerbated by changes to the revenue structure of the music industry at the macro-scale, where declining sales of recorded music in physical (CD) and digital (MP3) formats have increased the importance of the revenue derived from live performances. As the following musician points out, low barriers to entry are allowing musicians to flood the market for live music in Toronto and, as a result, finding steady and decent-paying gigs is increasingly difficult:

> It is common to not get paid or to have to pay to play, which of course makes no sense at all. Now (after three years) we are getting paid more consistently

but it is a maximum of $50 or $60 dollars divided amongst the members in the band. That does not even really cover the cost of equipment and rehearsal time. Usually it will buy you dinner for that night, and maybe the gas to get to the show.

(Drummer, March 2007)

Musicians also reported that too much 'buzz' was, in fact, a hindrance to productivity and the creative process. In the new era of independent production the free time once available to experiment creatively and indulge in the 'rock star' lifestyle has been lost. As a consequence, dedicated indie musicians in Toronto spoke of the danger of being sucked into projects and lifestyles that, in such a competitive climate, might derail their career goals. This musician, for example, saw his fledgling music career thwarted by a 'rock star fantasy' and cocaine addiction:

There is a partying lifestyle that comes with being a musician in a band ... There is late-night stuff, drinking, drugs ... you can get sucked into the party atmosphere as an entertainer ... I fell victim to it.

(Drummer, April 2007)

Lloyd (2006) indicates that musicians in Wicker Park, Chicago exhibited a strong connection to the drinking, drugs and partying that make up the 'rock star' lifestyle. The comments of this musician and others I spoke with, however, further illustrate that in order to succeed in the reconfigured landscape of independent music production, bohemian lifestyles, which are devoid of structure, responsibility, hard work and self-restraint, must be abandoned. The next section will demonstrate that the flight of musicians from the spaces of bohemia is not random, that a range of pull factors, including more affordable and artist-friendly space, better employment opportunities, greater control over their work/life balance and isolation from career-sabotaging temptations are attracting musicians to everyday spaces in the inner and outer suburbs of Toronto.

Creativity in everyday spaces

Just as the quest for cheap space helped to establish Toronto's original bohemian enclaves, as real-estate values in the downtown core continue to rise, affordable space is now attracting low-income musicians to the inner and outer suburbs. In Map 6.1, which shows the number of musicians per square kilometer in Toronto by place of residence, for example, it is clear that musicians are not exclusively clustered in the downtown core and can be found in many neighborhoods throughout the city.

As many of my participants complained about the costs of independent production and paying for equipment and advertising, finding cheap or even free space emerged as the most prominent pull factor for musicians. Moreover, as some of these musicians leave Toronto on tour for long stretches during the year, a further goal was to avoid paying high rents for unused space. The strategy of this

musicians/sq km
> 5
5.01 - 25
25.01 - 50
50.01 - 100
100.01 +
subway
streets

Produced by Matthew Talsma
Cultural Economy Lab - U of T

N

0   3   6   12 km

Source data: Statistics Canada 2006
Using NOCS code F033

Map 6.1 Number of musicians per square kilometer in Toronto by place of residence

musician was to move out of the bohemian inner-city neighborhood of Queen West in Toronto and relocate to Oakville, a suburban community within easy commuting range of Toronto.

> I used to live in Queen West, but in January I moved home with my parents, because I spend most of January and February on the road. I'm going to be gone for most of May, so I need to save money and stuff like that. So I've been commuting back and forth from Toronto to Oakville.
>
> (Singer songwriter and guitarist, March 2008)

This musician, who lives in a suburban community with no connection to hip hop culture demonstrates that even hip hop artists, who are described in the literature (Forman, 2000) as having the strongest connection to locally rooted scenes, value the affordability of space over its location:

> I think a lot of musicians ... are working 9 to 5, so they're maybe living in an apartment or a lot of times living at their parents' homes. It's like they're just trying to survive. But it's also funny because if you think of a lot of New York rappers, a lot of them don't live in Brooklyn and Queens like they say on their songs. Most of them live in New Jersey because that is where you are going to be able to afford a house. I'm up in the 'burbs, but I can still access the city by going to Scarborough.
>
> (Hip hop artist, April 2008)

Once again, unlike the bohemian notion of living and working within isolated artistic communities, we see that, as with networking, what really matters is the ability to access specific spaces in a 'just in time' fashion. Musicians can live anywhere as long as they can still get to their gigs, meetings and secondary jobs. In terms of accessibility, the best locations are described as being centrally located, often in the inner suburbs with good access to public transit:

> [After moving out of the downtown] we chose this site because it was the cheapest house we could get within the city limits. I didn't want to move to Whitby [outer suburb], because I would have spent too much time traveling ... Sometimes, however, there are a lot more work opportunities in the suburbs than there are in the downtown area of Toronto. There is so much competition, and they are fighting for fewer and fewer jobs. All the weddings, they are all in the suburbs anyway, banquet halls and golf courses. So as long as we can still access places like the Drake and the Rex [venues in downtown Toronto] that is all that matters. I think places like Mississauga and North York [inner suburbs] might be better places to live and work.
>
> (Pianist, July 2008)

These quotes indicate the decision-making processes of musicians and highlight the importance of finding affordable and accessible space. Musicians also reported being attracted to everyday suburban spaces because the built form is flexible and, therefore, more conducive to the creative process. In addition to needing enough space to store their equipment and hold rehearsals, for example, musicians also need to be able to make noise, often outside of the 9 to 5 work day. Interestingly, musicians also require silence to create and recharge from their hectic schedules. For these reasons some musicians prefer larger, more isolated spaces in the suburbs to small, crowded apartments in the city, with sleeping neighbors next door. Indeed, this musician moved to the outer suburb of Keswick after 20 years of living in downtown Toronto in order to make noise and concentrate on the creative process in complete silence:

> This is the twist, as a musician I make a lot of noise, which is bad enough, but I also need to live somewhere where there isn't a lot of noise, because I can't deal with that noise. I can't be creative with that noise. I need that silence to be effective and to focus on what I'm doing. I also need the peace and quiet just to rejuvenate myself from the stress of my working life.
>
> (Flute player and manager, July 2008)

In terms of the creative process itself, scholars including Jane Jacobs (1961) and Richard Florida (2002) have long argued that the high population density, short blocks and pedestrian access found in the downtown help to facilitate the interactions that support creativity, and that in contrast, post-war suburbs are the very definition of poor and uncreative urban form. Moreover, there is an assumption that the everyday spaces found in the suburbs lack the inspiration and 'authenticity' found in the downtown core. As this musician explains, however, the banal nature of suburbia itself served as creative inspiration:

> The title of our first album, *Parking Lot*, that sums it up right there ... The fact that I come from a place where there is nothing to do and a place where there is no music and where people are against what we are doing is my muse, it is the reason I write and started playing music in the first place ... I didn't belong so I invented my own thing to do, purely influenced by my surroundings.
>
> (Guitarist and singer, May 2007)

Furthermore, another musician balked at the assumption that creative or original music could not be produced in the suburbs and gave evidence of local bands in the outer suburb of Keswick creating meaningful new musical forms:

> We are a Keswick band and there is definitely a Keswick sound ... There is this whole thing of art rock being mixed in with emo music in Keswick, that is sort of the sound which is coming out of there right now. Which is really kind of interesting.
>
> (Drummer, February 2006)

Despite vernacular and apparently sterile aesthetics, therefore, suburbs such as Keswick can, in their own way, act as intersections of new ideas, styles and creativity. Perhaps even more important than the ability to facilitate the creative process and provide affordable and flexible space, Toronto's suburbs allow struggling musicians to sustain their creative passions by providing better employment opportunities. Although Toronto's downtown core is saturated with an oversupply of musicians, which is limiting the amount of paid employment, the markets for live music in many outer suburbs remain untapped. For example, this musician explained that playing shows in the outer suburbs and smaller towns in the periphery often generates better fan attendance and pay checks precisely because the market is not saturated with musicians and other entertainment alternatives:

> The music scene in the slightly less populated areas north of the city is getting to be really good ... In the northern areas, the kids have less to do, there are fewer entertainment options for them, in Toronto there are one million things to do, so if there is a live band, those kids are gonna go, so you can sell tickets easier. The highest turnouts to any of our shows have all been in Newmarket and Keswick [both outer suburbs of Toronto].
>
> (Drummer, February 2006)

As these findings suggest, musicians make spatial choices based on their own unique set of criteria, with the goal of achieving the optimum balance between a range of factors. Most notably, these include the affordability, accessibility and artist-friendliness of the physical space and, crucially, the availability of paid work.

Conclusion

The existing accounts characterize bohemia as both highly concentrated and uniform across time and space. More broadly, the literature constructs a stark dichotomy between the downtown core and the surrounding suburbs. The core, which is said to radiate 'authentic', 'alternative' and stimulating energy, is regarded as the spiritual home of the creative process and thus the domain of artists, musicians and other bohemians. In contrast, the suburbs are depicted as sterile, banal and vernacular spaces populated by 'square' professionals who are anything but creative. As I have argued, however, the demands of independent production in the digital era are forcing some musicians to abandon bohemia. I have demonstrated that the lifestyles of indie musicians are moving closer to those of mainstream professionals and that a range of push and pull factors, including the quest for affordable space and employment opportunities, have resulted in some musicians leaving the core and relocating to spaces in the inner and outer suburbs. These everyday suburban spaces, parents' houses and basement apartments, for example, are far more than cheap containers of creative activity and serve to support and catalyze the creative process in their own right. To recapitulate, I presented examples of musicians drawing inspiration from the ordinary routines found in suburbia and reworking these conventions in creative ways. Moreover, I discussed how musicians in the outer suburb of Keswick are appropriating, recombining and infusing sonic styles from the downtown core with their own ideas to create new, hybrid forms of music. While it is clear that some musicians are being forced to abandon bohemia, others are making this transition of their own free will. As the trends toward making spaces of creativity in everyday realms and adopting 'square' lifestyles accelerate, important questions about the true nature of creativity and what it means to be an artist in the digital age are raised. The results of this micro-scale case study lend further credence to the mounting criticism directed at academic inquiry and government policies that privilege the visible clusters of creative activity in the downtowns of cities and neglect their invisible counterparts in everyday and vernacular spaces.

7 Mapping vernacular creativity

The extent and diversity of rural festivals in Australia

Chris Gibson, Chris Brennan-Horley and Jim Walmsley

Introduction

The idea that creativity is vital to regional economies has been increasingly debated in Australia, as elsewhere. Although creativity has been taken more seriously by governments, it has often been 'folded into' normative ideas of market-led place competition, with biases towards commodifiable forms of creativity (that produce copyright/content) and urban, middle-class neighbourhoods and aesthetics (see Gibson, 2009 for extended critique).

In reaction to these observations of the particularities of creativity policy in Australia, a central research concern in recent years has been to expose such biases as embodied practices of knowledge construction (Gibson and Klocker, 2004); to trace the manner in which such knowledges about creativity are produced and circulate through various 'scenes' and sites of knowledge re-production (academic, technical, policy etc. – Kong *et al.*, 2006; Barnes *et al.*, 2006); and to problematise formulaic visions of what constitutes creative industries, by exploring 'everyday' creative expressions in suburban, rural and remote settings where boundaries between 'amateur' and 'professional' creativity may be porous. This is an area of emerging strength in geography – where researchers have been particularly keen to unsettle assumptions about the automatic links between creativity and cosmopolitan and large urban settings (Gibson, 2002; Bell and Jayne, 2006).

In the first part of this chapter we overview the manner in which normative discourses of creativity have infused policy talk. In the second part, the chapter draws on one project which has sought to map the extent and diversity of creativity in rural areas, specifically, community festivals held in rural parts of three Australian states (Tasmania, New South Wales (NSW), Victoria). We discuss in detail one festival: the Elvis Revival festival in the country town of Parkes. Such events are rarely taken seriously by local governments and development promoters in rural regions, who favour 'traditional' regional industries (like mining and agriculture). Nor are they noticed much by proponents of 'new' or high-tech creative industries, usually based in cities, that are highly visible and that have links to corporate interests and promise export growth (e.g. film, digital design). Rural community festivals are also often ignored by urban-based arts policy makers – because they either are considered barely 'creative' (in an artistic sense) or are associated with rural working-class cultures, considered redneck or 'hick', that disturb or offend urban cosmopolitan fancies. This chapter accordingly seeks to show that rural creativity is widespread, articulated via vernacular, grass-roots cultures, and is intimately connected to ideas of community sustainability and vitality. We argue that festivals ought to be taken seriously, but that this requires a shift in the tenor, meaning and class politics of regional development discourses.

Normative creativity in Australia

In a paper co-written by one author with Lily Kong (Gibson and Kong, 2005), a critique was outlined of the manner in which a common policy imperative has often been to make generalisations about the creative or cultural economy as a transformative component of total economic activities in places, such that it can be considered as a whole-of-economy phenomenon. Such generalisations become normative, where meanings for creativity coalesce around singular, definitive policy interpretations. If 'normative creative economy' could be distilled into a single script, it would probably look as follows:

> Contemporary capitalism is characterised by more recently dominant forms of accumulation, based on flexible production, the commodification of culture and the injection of symbolic 'content' into all commodity production.
>
> Some places do better than others from this: those that have highly skilled, creative, innovative, adaptive workforces, as well as sophisticated telecommunications infrastructures, interesting and diverse populations, relatively low levels of government interference in regulating access to markets, as well as lifestyle attractions, restaurants and arts institutions to attract a new 'creative class' (a social segment who avidly produce and consume the output of creative industries).
>
> In order to compete in the new cultural economy, places ought to implement particular policy initiatives: encourage cultural industry clusters, incubate learning and knowledge economies, maximise networks with other successful places and companies, value and reward innovation, and aggressively campaign to attract the 'creative class' as residents.

In the 1990s and early 2000s, this normative approach proved attractive in several cities in the Asia-Pacific, particularly those such as Auckland, Sydney, Hong Kong and Singapore – important regional cities with already established national broadcasting, arts and cultural industries, but with aspirations for 'world city' status (see Kong *et al.*, 2006 for discussion). In Australia, creativity has been conceptualised and stylised by the propagation of a particular regional development template, mostly neoliberal in flavour, orientated to finding market-based solutions to problems of uneven distribution of economic activity and wealth. Creativity is a key concept of interest for governments because of its polysemic nature: as well as defining a state of expressive humanity (an omnipresent human act with seemingly unarguable, positive overtones about freedom, vitality and excitement), creativity is also capable of being woven into governmental agendas that problematise – and in turn offer up solutions to – particular social groups and places

(Bill, 2008). Problem locations and social groups (in Australia, regions with high unemployment, declining populations, or large proportions of Aboriginal people) are identified by governments as being in need of transformation. Creativity has proven to be one concept offering hope for such transformations.

One example was the 2002 replication in Australia of Richard Florida's creative index benchmarking of regions (National Economics, 2002). In it, an attempt was made to apply Florida's (2002) 'creativity index' to Australian regions through a statistical measurement of employment in creative industries, cultural diversity and patent registration. Of the 64 regions analysed across the country, the lowest dozen were all sparsely populated inland agricultural regions in NSW, Victoria, Queensland and South Australia. 'Winning' regions were all in the central areas of state capital cities – and predominantly in Sydney and Melbourne (Australia's two biggest cities, with around 5 and 4 million people each, respectively). Not only did non-metropolitan and poor regions 'lose' on the creativity index: in the report's text, poor regions are described as 'problem' locations that 'lack' creativity, while actual systemic causes of economic disparity (uneven distribution of resources, capitalist modes of production, impacts of restructuring etc.) are not mentioned. In benchmarking exercises such as the *State of the Regions* report, rural, remote and poorer industrial regions are portrayed as problem places because they lack innovation and are thus in need of policy fixes (see Gibson and Klocker, 2005 for extended critique). But rather than fund better services, invest in training and job creation or address the underlying root causes of socio-spatial disparities (which ultimately lie in the contradictions of the capitalist space economy), creativity is offered instead as a human condition to nurture and promote. The solution to the problems of disadvantage is thus imagined as a cultural transition of individuals from welfare-dependent, problem subjects into 'entrepreneurs' and 'consumers', the 'idealised companions' of the neoliberal state: 'liberated, independent and competitive' (Peck, 2004: 395). Creativity is increasingly popular because it is positioned at the core of innovation, invention and enterprise culture, and because its industrial outputs – most obviously those in the creative industries such as film, fashion and music – are trendy consumer objects at the basis of sectors with growing workforces and export potential.

Glossy brochures have since been produced as cities have funded creative city planning strategies, and talk of 'innovation' has infused corporate, university and government circles. How this has played out in actual policy terms is only now being realised – and the picture is of somewhat haphazard and frequently uninspiring policy making. On the one hand, there is a tendency towards promoting corporatised creativity (the film industry is often the first to be envisioned as desirable); on the other, arguments about the value of the institutionalised arts are repeated, and reorientated, away from the importance of public subsidy, to the anchoring role that arts institutions can play in the creative life – and spatial planning visions – of cities and regions. Frequently, policies aimed at fostering creativity are tailored to the imagined capacities of particular social groups and places: thus, in remote Australia, Aboriginal people are encouraged to develop traditional visual art and performances for tourist and export markets, as a succession of employment and empowerment strategies target creative industries as a 'fix' for unemployment and social deprivation (Gibson, 2009). In big cities, by contrast, local governments fund the construction of specially planned cultural districts (in an increasingly formulaic blueprint fashion, where regional art galleries, entertainment centres and exhibition spaces are clustered together) or the hyping of inner-city gentrification and café society in main street or neighbourhood place promotion campaigns (with few connections to creative work and production, per se). The motivation is to embrace creativity without jettisoning previous governmental objectives; the effect is the maintenance of a status quo – and a decidedly middle-class aesthetic throughout.

How might this mask other diverse, alternate and vernacular understandings of 'creativity'? Recognition of these other creativities requires a careful critique of, and commitment to move away from, doctrinaire positions about regional culture, market forces and place competition (Gibson-Graham, 2008). It also requires attention to be paid to the sorts of creativities enacted beyond familiar industries and places.

The rural festivals project

The remainder of this chapter builds on this desire to pay attention to the creativities enacted beyond familiar industries and places by reporting from a research project undertaken into the social and economic dimensions of cultural festivals in non-metropolitan parts of Australia. The research project, running from 2005 to 2008, involved John Connell and Gordon Waitt as collaborators and was funded by the Australian Research Council (DP0560032, 'Rural festivals as regeneration strategies'). It sought to interrogate the manner in which festivals of all types contribute to the changing social and economic life of rural and regional places.

The project has had three phases. Phase one was the compilation of a database of cultural festivals outside the capital cities of three Australian states – New South Wales, Victoria and Tasmania – regardless of type or location. The aim was to locate and list as many festivals as possible occurring in a 12-month period (March 2006 to March 2007) in order to capture as accurately as possible the complete picture of the breadth and diversity of rural festivals. Across the three states, the details of 2,856 festivals were recorded.

The second phase of the research was a postal survey sent to a subset of the above festivals (those for which we had full contact addresses and the name of an organiser). The survey consisted of seven pages of questions, ranging from the sponsorship, income, aims and organisational structure of the festival, to Likert-scale questions gauging the responses of organisers to a set list of questions about tourism, regional identity, relationships with host communities and impacts of drought. Valid, completed surveys were returned by 480 festivals.

Creativity in diversity

The results from our database and survey indicate the breadth and diversity of festivals held in rural areas (Table 7.1 and Map 7.1). This provides evidence of lateral and innovative thinking within communities that are assessing their resources and managing to stage successful events. The most common were sporting, community, agricultural and music festivals – which, combined, made up three-quarters of all cultural festivals in non-metropolitan areas. Even though these festival types dominated, within these there was further diversity: 'community' festivals covered everything from Grafton's historic Jacaranda Festival (named after the town's signature tree) to Kurrajong's Scarecrow Festival, Nimbin's Mardi Grass (a marijuana pro-legalisation festival), Ballarat's Stuffest Youth Festival (which presumably is, 'you know, like, about stuff …'), Ettalong's Psychic Festival, Tumut's Festival of the Falling Leaf, Myrtleford's curiously amalgamated Tobacco, Hops and Timber Festival, and Thoona's even more bizarre Wheelie Bin Latin American Festival (an ostensibly music and dance festival which culminates in a billy-cart type race down the town's main street, of vehicles made by entrants from large, domestic, wheeled rubbish bins). Similarly varied were sports festivals, covering everything from fishing to billy carts, cycling, pigeon racing, hang gliding, dragon-boat racing and camp drafting, an Australian sport involving mounted horse riders demonstrating droving skills by navigating individual cattle through gates and obstacles.

Table 7.1 Numbers of festivals, by type, Tasmania, Victoria and NSW, 2007

Type of festival	TAS	VIC	NSW	TOTAL**	% of total
Sport	86	485	488	1,059	36.5
Community	45	216	175	436	15.0
Agriculture	19	146	215	380	13.1
Music	13	116	159	288	9.9
Arts	12	73	82	167	5.8
Other*	7	87	71	165	5.7
Food	10	53	67	130	4.5
Wine	7	49	32	88	3.0
Gardening	20	43	14	77	2.7
Culture	2	21	11	34	1.2
Environment	1	8	12	21	0.7
Heritage/historic	4	8	7	19	0.7
Children/youth	0	10	5	15	0.5
Christmas/New Year	0	10	2	12	0.4
Total	226	1,325	1,340	2,891	100.0

Source: ARC Festivals database, 2007.

Notes
* The 'other' category includes small numbers of the following festival types: Lifestyle, Outdoor, Science, Religious, Seniors, Innovation, Education, Animals and Pets, Beer, Cars, Collectables, Craft, Air Shows, Dance, Theatre, Gay and Lesbian, Indigenous, and New Age.
** The total for this table is slightly more than the total number of festivals in the database, due to counting of some festivals in more than one category. This occurred when separating categories proved impossible (for example, for 'food and wine festivals').

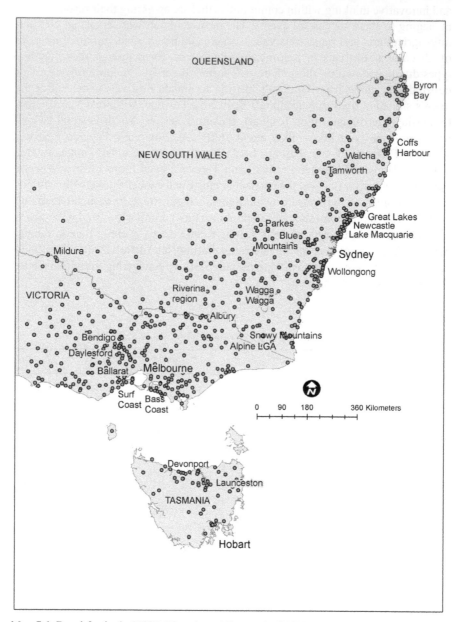

Map 7.1 Rural festivals, NSW, Victoria and Tasmania (2007)

Among music and arts festivals there was some diversity, although certain festival types dominated (Table 7.2). In music, country, jazz, folk and blues festivals counted for over half of all music festivals (Gibson, 2007) – far outweighing styles such as rock that are more commercial or lucrative in the wider retail market for recorded music. In the sphere of the arts, film festivals, generic 'arts festivals', visual arts (usually painting, occasionally photography and only once sculpture) and art and craft festivals dominated. Overall, music and arts festivals could be considered more 'vernacular' than 'elite' with regard to class stereotypes: sewing and quilting festivals were as common as opera festivals; country music was more prevalent than jazz (although the latter is remarkably widespread, given that it is otherwise a statistically small niche of the recorded music market), and also present were other 'roots' music styles such as bluegrass and folk. When

Table 7.2 Arts and music festivals, by type, Tasmania, Victoria and New South Wales, 2007

Type of music festival	Total number*	Percentage of all music festivals*	Type of arts festival	Total number	Percentage of all arts festivals
General music/not specified	90	31.3	Film	45	26.9
Country	56	19.4	General arts	43	25.7
Jazz	50	17.4	Visual art	30	18.0
Folk	31	10.8	Art and craft	10	6.0
Blues	20	6.9	Drama	9	5.4
Rock and/or pop	11	3.8	Sewing/quilting	7	4.2
Other**	11	3.8	Writers	5	3.0
Classical	9	3.1	Dance	3	1.8
Opera	8	2.8	Performing arts	3	1.8
Bluegrass	4	1.4	Poetry	3	1.8
Rock 'n' roll	4	1.4	Youth arts	3	1.8
Scottish	4	1.4	Comedy	2	1.2
Alternative rock	3	1.0	Literature	2	1.2
Indigenous	2	0.7	Children's literature	1	0.6
Irish	2	0.7	Puppetry	1	0.6
Rhythm and blues	2	0.7			
World music	2	0.7			
Total	298	–	Total	167	100.0

Source: ARC Festivals Project database, 2007.

Notes
* The total number of music festival numbers across the various categories in this table is greater than the actual number of festivals in the database due to about 10 per cent of festivals listing multiple genres (e.g. jazz and blues, folk and country). In such cases genres were counted separately. Hence figures in the 'percentage of all music festivals' column equal more than 100 per cent.
** The 'other' category includes single music festivals of the following types: metal, reggae, dance music, rockabilly, Celtic, Christian, German, hip hop, Indian, contemporary and Latin.

measured by numbers of attendees, music festivals were among the largest of all festivals, including Tamworth's annual country music festival, Tweed Heads' 'Wintersun' Rock and Roll/1950s nostalgia festival, Goulburn's Blues Festival, Lorne's 'alternative' Falls Music Festival, Byron Bay's East Coast Blues Festival and Splendour in the Grass. Audience sizes at these were in the range of 15,000 to 100,000 people. But these were exceptions rather than the norm: the average attendance size was 7,000; and 30 per cent of all festivals surveyed had audiences of under 1,000 people.

It would be wrong to automatically interpret this abundance of rural festivals as utopian – we documented plenty of instances of fractious host communities, and tensions between perceived 'insiders' and 'outsiders' within individual festivals (see Gorman-Murray *et al.*, 2008). Festivals always have the capacity to selectively seek and represent some elements of local cultures and identities while evicting others, thus intensifying social exclusion – inadvertently or otherwise (Atkinson and Laurier, 1998; Gibson and Davidson, 2004). Yet, political economic critiques often levelled at festivals elsewhere (and particularly in large cities) scarcely seemed appropriate for the circumstances we documented in rural Australia: rarely were festivals commercial (74 per cent were run by non-profit organisations; only 3 per cent were run by profit-seeking companies) or incorporated into formal economic planning or branding strategies (24 per cent of all surveyed festivals). With little reciprocal interest from big business or the powerful arms of the state, they were rarely part of civic boosterism (cf. Boyle, 1997). Nor have they been swept up in the broader, entrepreneurial neoliberalisation of contemporary government in the manner of creative industries in cities mentioned in our introduction (cf. Hall, 1997; 2006; Muñoz, 2006).

Instead, the stated aims of festivals were more often than not linked to the pastimes, passions or pursuits of the individuals on organising committees, or to socially or culturally orientated ends such as building community, rather than as income-generating ventures (Table 7.3). Indeed, of all categories of festival aims, 'to make money' and 'to increase regional income' were the two rarest responses (recorded in only 5 per cent of cases, combined). In short, rural festivals in Australia are not particularly capitalist in nature. They are, more accurately, hybrid economic forms: grass-roots affairs run by charities or small committees of interested individuals brought together by shared hobbies or interests. They are part of the community economy, but they are also linked with elements of the government sector, the private sector, an exchange/gift economy (quid pro quo arrangements between local businesses and festival-organising committees are very common), and the informal sector. Some local businesses – pubs, motels and cafés – usually benefit, while others do not. There is also a network of itinerant stallholders who travel from festival to festival in an annual circuit, earning a living selling food, arts and craft, clothing or bric-a-brac (the average number of stalls at festivals was 67). Considerable formal employment is generated: we estimated that 176,500 full-time and part-time jobs were created directly in the planning and operation of cultural festivals in regional Australia in 2007, based on survey returns – equivalent to total employment in these same regions in agriculture (see Gibson *et al.* forthcoming). Yet overwhelmingly they rely on volunteers:

Table 7.3 Aims of surveyed festivals

Aim	No. of festivals*	% of festivals surveyed
To promote a place/theme/activity	137	28.5
To show(case) a place/theme/activity	86	17.9
To build community	75	15.6
To compete	75	15.6
To entertain	65	13.5
To foster/encourage	63	13.1
To celebrate	44	9.2
To fundraise	41	8.5
To educate	21	4.4
To make money	12	2.5
To increase regional income	12	2.5

Source: ARC Festivals Project survey, 2007.

Note
* The total number of festivals by aim is greater than the total number of surveys received. Festivals that recorded more than one aim are counted for each of these records.

19.2 days were spent by the average volunteer assisting their festival during its planning phase, and 5.7 days on average assisting during the running of the event at time of operation. Across the 480 surveys this constituted the equivalent of over 8,600 days (or 23 years' worth of labour) when adding up the work done by *the average volunteer* across all festivals in that given calendar year.

In some cases we documented, it could be true to say that festivals served the interests of the elite (at least, at some classical music festivals) and tended towards 'safe' culture (see Gibson and Connell, 2009); but mostly they were accessible, free (unticketed), 'earthy' and sometimes outright bizarre in character. Unlike at festivals elsewhere (see Waitt, 2008), rarely were they instruments for the powerful within local communities to feather their own nests and entrench their local political and commercial positions. Indeed, in many cases a much bigger issue was how to maintain sufficient interest from newcomers and younger generations in running the event after older, original organisers had retired (or died). Demographic changes afflicting most of rural Australia (particularly out-migration and ageing) impacted profoundly on their longevity. Yet despite these challenges, the picture created by our attempts to document the breadth and diversity of rural festivals in Australia is of a particular kind of vernacular creativity: less concerned with the production of material goods or intellectual property, and more about communities with patchy skills in event management being inspired to come together for fun or to celebrate a shared passion, and subsequently marshalling often limited resources within their towns and villages in order to make a festival happen. It illustrates a form of vernacular creativity more widely present in rural communities: creativity through ingenuity and action, rather than in artistic output per se.

Creativity in imitation

The third, and concurrent, phase of this research involved qualitative ethnographic work carried out at selected festivals, and with the approval and participation of their organisers. One of these was a music festival: the Parkes Elvis Revival Festival, held every January on the anniversary of Elvis Presley's birthday, in Parkes in the NSW sheep/wheat belt. Parkes is a small country town of 10,000 residents, 350 kilometres from Sydney (Map 7.2). Like many other inland country towns, it had lost population (4 per cent between 1996 and 2001), had higher than average unemployment rates and low levels of participation in the labour force (43 per cent of total population), particularly as its population became dominated by those of retirement age (ABS, 2001). It was essentially a service centre for a rural agricultural region located in Australia's sheep/wheat belt. Plagued by drought since the late 1990s, farmers sought to diversify into other crops such as canola, but the precarious nature of wheat exports still troubled the town. Other than its historic radio telescope ('The Dish'), a vital link in the 1969 Apollo moon landing (which was, in 2000, the subject of a popular Australian feature film of the same name), Parkes had little in the way of visitor attractions. It was once a major hub; however, rationalisation of the NSW rail network (which would mean further job losses for the town) reduced visitor through-traffic. In the 2002 *State of the Regions* report its wider region (the NSW Central West) was ranked 51st out of 64 across Australia on the creativity index.

This depiction – of rural decline, ageing and lack of creativity – has been seriously undermined in recent years by the manner in which grass-roots creative activities have coalesced in Parkes around its Elvis Presley Revival Festival. Far from being a formal creative place strategy, the emergence of the Elvis Presley Revival Festival in the early 1990s was entirely the result of a chance local whim, when local business owners, devoted to the memory of Elvis, proposed the idea to local council members, as recalled by committee member Neville 'Elvis' Lennox:

> It was Bob and Anne Steel up at Gracelands restaurant. They're big Elvis fans and they own the restaurant. They were just having a bit of a talk to the right people at the right time, at one of their functions. They were councillors and they said, 'Well there's nothing going on, nothing celebrated that time of year. Elvis's birthday's the eighth. Come along to the next council meeting, we'll put it to the board.' It just evolved from there.
>
> (Personal communication, 2004)

Parkes happened to have a restaurant called Graceland and a small group of committed fans willing to organise an event. This suited the pragmatic aims of the local council of the time to improve summer tourism, as conveyed by Parkes Shire tourism manager Kelly Atkinson:

> The tourism board and council together recognised that January was a very quiet time of year. They were trying to introduce more events onto the calendar

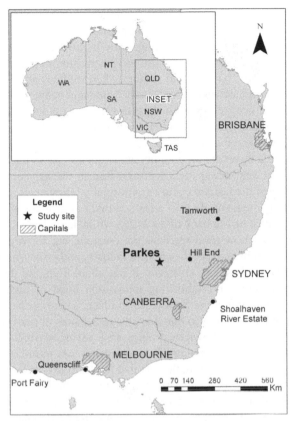

Map 7.2 Location map, Parkes

having identified tourism as a key market to target. In consultation with local businesses such as Gracelands they came up with the idea to have a birthday party for Elvis. It started as an idea to try and turn that low season around and invent a festival that would curb that.

(Personal communication, 2004)

An Elvis Revival committee was subsequently formed, and in 1992 what was essentially a very small group of local fans decided to stage Australia's first Elvis festival. The first Festival was held in January 1993, coinciding with Elvis's birthday. It attracted 500 people from as far as Adelaide, Melbourne and Sydney, and set the theme for those that followed: a festival which at the core celebrated the life and music of a dead rock star, through imitation and themed activities.

Since then the festival has been staged annually – and despite some years when struggles to get sufficient volunteers hampered organisation and promotion, it is now a professionally run major event for the region, attracting between 6,000 and 8,000 visitors to Parkes and generating over A$4 million (£1.8 million) in direct visitor expenditure in the town (Brennan-Horley *et al.*, 2007; Stanes *et al.*, 2007; Ruting and Li, 2008). The festival attracts certain demographic groups: 64 per cent were between the ages of 45 and 65, with another 12 per cent over the age of 65 (probably unsurprising – as this demographic is most likely to have had Elvis fans when he was still alive and performing). Increasing in presence is also another demographic group: young urban visitors, often students and fans of popular culture, who enjoy the festival for its 'kitsch' value and sense of absurd fun (see below). Only 15 per cent of the festival goers are from the local region. Over a third usually come from Sydney; another third come from other parts of non-metropolitan NSW; and the remainder come from interstate (with virtually no international visitors). Media coverage on commercial television news programs and in metropolitan and national newspapers has increased in recent years (e.g. Minus, 2007) – particularly as the festival occurs in a normally 'slow news' period of the year. The overarching media angle is of a festival that has grown from tiny origins to national significance on the back of a kitschy, whimsical idea.

And this is the point about how rural festivals are creative, but in ways that contrast the assumed blueprint in much of the creative industry and policy-making literature: at the Elvis Revival Festival, creativity is not about high-end design, luring metropolitan film makers or establishing innovation hubs of high-tech SMEs – strategies repeated ad infinitum from place to place. There are no truly 'original' products made in Parkes, or even home-grown rock bands or performers in the style of Elvis who produce their own recordings. Instead, creativity is about a quirky idea followed through with a range of activities that mix mimicry with nostalgia, humour with retro kitsch. Indeed, the fact that the festival is about Elvis Presley – a performer who had never visited Australia, let alone Parkes – negates any existence of the sort of 'authenticity' often sought after in cultural tourism and intellectual property-generating creative industries. As one participant in the 2007 festival argued:

> You've got Elvis wine, Elvis beer, Elvis tooth brushes, there's heaps of stuff; it's really tacky ... like Louisiana mud ... the tackier it is the better it is ... I mean people are buying 45-foot Elvis rugs [laughs]. Which is classic ... classic behaviour at a festival that's focused on one thing ... it doesn't matter any more, the details are irrelevant. People consume all this memorabilia because people are in the spirit of it and that's what a festival does, it changes your behaviour.

The festival begins on the Friday night of the weekend closest to Elvis's birthday (8 January 1935), usually with dinner and various forms of Elvis enter-tainment (with all participants encouraged to dress in appropriate annual themes: cowboy, speedway, Hawaiiana – usually linked to Elvis movies). Saturday sees the street parade of vintage cars and motorbikes (and vintage Elvis impersona-tors), with market stalls (ranging from memorabilia – rarely 'real' – to country handicrafts) in the town's main park area (Figure 7.1). The park is the venue for sound- and look-alike competitions – Elvis, Priscilla, Lisa Marie and Junior Elvis – and the day concludes with a feature performance by a touring 'professional'

Elvis impersonator. The highlights of the Sunday are the highly attended Gospel Church Service, the Elvis celebrants' renewal of wedding vows of hundreds of festival participants, and the unveiling of a new plaque on the Elvis Wall (at the park) to commemorate another 'legend' of Australian rock 'n' roll music (often the previous night's top-billing performer). The wall itself surrounds gates that are a replica of the gates of Presley's Graceland mansion in Memphis. A talent contest with more diverse themes brings the festival to an end as most visitors return over considerable distances. A special train (the Elvis Express) runs from Sydney (Figure 7.2), with the support of NSW Railways and the state tourism promotional authority. On some occasions Elvis movies have been shown and the local lawn bowling club has urged visitors to 'kick off your blue suede shoes' and have a game. An Elvis celebrant can be made available for couples to marry or renew marriage vows during the weekend, a handful of buskers occupy street corners, and the private collection of memorabilia of Elvis Lennox – with a pink Cadillac parked in the driveway – is open to all visitors.

This is vernacular creativity less about producing things for sale (although one stallholder interviewed at the 2009 festival described how she made her own Elvis-themed knick-knacks for sale and travelled over 1,000 km to attend the festival), than about dressing up, acting up and losing one's inhibitions – creativity in costume design, in personal transformation and release, rather than overtly market-orientated activity. The following quotes from 2007 festival participants suggest how participants, particularly those in Elvis jumpsuits, enact this peculiar kind of personal, performative creativity:

> We were talking about what happens when you put on an Elvis suit, it changes you. It changes your behaviour, everything changes … it changes the people around you. Everyone that notices you, talks to you and yells out to you. It brings an uninhibited kind of feeling people just all come together. They let loose, they have an excuse … It's a small rural town socially acceptable way to play dress ups.

> You can act out a bit more, you get to meet more people through it, they seem to come up to you … I mean whereas if you were casually dressed they wouldn't give you a second glance. It gives you a bit of adrenaline I suppose; you know … you feel a bit like the King probably would have felt you know.

> The other thing we noticed when we were walking down the main street last night, everyone was singing, everyone was friendly and it sort of unusual because there's no inhibitions and everyone is 'yahooing' and having a few beers, which was great.

> It is very different to your everyday life because when they look like Elvis they are posers, and you stop them to take pictures of them and they are ready to pose for you, and when they take their wigs off they're just normal people. It really allows them to hide behind the mask and act out … we could have

Figure 7.1 (this page and opposite) The Parkes Elvis Revival Festival, 2007

gone and put a couple of wigs and we could have walked down the main street and started singing and I wouldn't have felt bad at all because no one would have known who we were or cared.

These comments were reflected in overall results of a survey of over 300 visitors undertaken at the 2008 festival: 11 per cent of all visitors were also performers (a comparatively high percentage for music festivals); and after music and entertainment (which unsurprisingly were the most commonly cited reasons for visiting the festival), the main motivations for visitors attending the festival were 'social interaction' (24 per cent) and 'escape from the everyday routines' (13.1 per cent). Audience participation in mass imitation (of Elvis, Priscilla *et al.*) is what

constitutes creativity here. Several Priscillas sewed their own costumes (and those of their accompanying Elvises) – although men tended to hire their jumpsuits rather than make them – and impersonators are rewarded for their likenesses, but also for their showmanship, exuberance or particularly quirky 'twist' on the Elvis formula (e.g. fat Elvis, old Elvis, gyrating dance moves). For one local fan, the art of imitation even encouraged him to take on the icon's name:

> I prefer Elvis to Neville, me original first name. After the first two years of competition here in the look-alikes – I won that in 93, 94 – and walking up the street or down the street, you hear people yell out across the street at ya: 'g'day Elvis' and that. And I said, 'ya know, that would be an idea'. So I put

Figure 7.2 The Elvis Express

it to me mother, asked her permission to do so and she said 'you go ahead and do with it what you want.' And I said, 'thank you very much'. Paid 75 dollars and had it legally changed to Elvis.

(Personal communications, 2004 and 2007)

At Parkes, the self was literally re-created – perhaps the ultimate example of vernacular creativity at work.

Conclusion

From cursory consideration here of the extent and impact of rural cultural festivals, and from a vignette of one particular event, it is clear that rural creativity is prevalent, that it is catalysed in a range of places and occasions – some of which are large, tourist orientated and part of coordinated regional marketing and development strategies, but most of which are small, informal and connected to quirky themes, charity goals and personal leisure interests. Diversity defines rural creativity, unlike the normative creative city planning agenda: from small community events celebrating migrant heritage to quirky, specialist offerings such as scarecrow festivals, pumpkin bakes, goanna-pulling contests, 'ute musters' and chocolate fairs. In the staging of the festivals themselves creativity is evident in their uniqueness, and in the lateral thinking, open-mindedness and humour they embody. Even in imitation – as at Parkes – rural life is eminently creative.

What might this story mean for how we theorise the geography and significance of creativity? In the first instance it means that assumptions about 'where creativity is' demand critical attention (see too, Brennan-Horley and Gibson, forthcoming). That creativity might be omnipresent is fairly obvious – but yet as Pratt (2004), Bell and Jayne (2006) and Brennan-Horley and Gibson (forthcoming) have all suggested, there persists a certain kind of geographical illiteracy in much creativity-inspired research and policy making. Cities remain the sites of most research, and the assumption is that creativity acts centripetally to agglomerate activities in webs of closely related firms and entrepreneurs. We have shown here, through documenting the breadth and location of rural festivals, that creativity can likewise reveal itself as dispersed, atomised and hybridised: at the edges of the community, government and informal sectors of the economy.

This story also suggests that the manner in which people participate in creative activities demands more consideration. Creativity is not just about firm locations or outputs, or commodifiable innovation that is transformed into exportable product or intellectual property. Creativity is deeply present in how people engage with each other, assess their access to resources and the situations facing them and their regions, and build on nascent ideas in interesting and innovative ways (Eversole 2005). Such articulations of creativity are perhaps far closer to the spirit of the word than the more thoroughly neoliberalised, policy-driven script. In cases such as festivals, creativity enables rural people to transcend everyday life, generating more interesting and varied forms of work (even if temporary), attracting notoriety for regional cultures and contributing to the vitality and diversity of places.

Recognition of this more humanistic understanding of creativity does not mean we need to abandon thinking about economic significance, nor take less seriously the potential of creativity for alternative pathways to regional development. From the 2,800+ festivals in the rural festivals project we estimate that, cumulatively, these alone bring over A\$1 billion (£450 million) in direct income to the places in which they are hosted, largely without corporate support, and mostly driven by non-profit organisations and local committees. And yet rural festivals rarely figure in the policy framing of regional economic futures – even when regional development policy making has sought to engage with the idea of the creative economy. We would argue that this reveals the manner in which economic knowledges about places are selectively and partially produced. In policy making, creativity appears to matter most when it is seen through particular political-economic lenses that privilege visible projects and industries, export markets and private sector investment. We hope that our documentation of rural festivals across southeastern Australia goes some way to redressing this rather skewed way of understanding how people constitute an economy.

8 Imagining the spatialities of music production

The co-constitution of creative clusters and networks

Bas van Heur

Introduction

The important role played by spatially concentrated forms of cultural production is constantly proclaimed and assumed – not only by policy makers in their quest to develop new spaces of economic development, but also by artists, journalists and other cultural entrepreneurs active within their respective cultural scenes and milieus. The aim of this chapter is to relativize some of these claims by focusing on the mutual intertwinement of (spatially concentrated) creative clusters and (spatially distributed) creative networks and, by doing so, to problematize some of the core assumptions of cluster approaches. The empirical focus will be on what is usually called electronic music or club cultures, since this particular form of contemporary music production allows for a rich illustration of the ways in which clusters and networks are intertwined. The main argument will be that electronic music clusters and networks co-constitute each other at all times. Clusters, in other words, are dependent on networks for their emergence and development, but networks also rely on clusters for their own reproduction and transformation. It is this intertwinement that enables non-economic impulses to shape the very formation and characteristics of particular spaces of music production.

The outline of this chapter largely follows the broader argument of this volume, as described by the editors in their introduction. After providing a preliminary definition of electronic music and drawing on completed research on London and Berlin, the chapter briefly addresses the role of creative cluster policies in these two cities. This way of imagining the spatialities of cultural production is criticized for its economic bias and is then contrasted with the actual spatial dynamics of electronic music production in and through London and Berlin. In describing these dynamics, multiple values and determinations beyond the economic are highlighted. Finally, the last section summarizes the argument and points to the role of socio-spatial imaginaries in representing, regulating as well as transforming cultural production.

Electronic music production – a preliminary description

The notion of electronic music in this chapter functions as shorthand for those forms of music production that rely on electrically produced sounds, created with computers, virtual instruments (software), synthesizers, samplers and other tools that can only function when connected to the electrical circuit. From this perspective, I also include record players and the use of record players by DJs as belonging to the category of electronic music, since, first of all, the recorded music played is often electrically produced and, second, the use of the electrically powered record player as an instrument is central to the aesthetics of electronic music. This definition is a pragmatic one, however, since, even on the level of sound sources, there are many hybrids. Electro-acoustic music, for example, uses sounds from the outside world (the weather, the built environment, household objects, but also acoustic instruments) and is thus not properly electronic, but these sounds are often electrically modified to such an extent that I would consider this music to be electronic music. Similarly, voices or samples from traditional instruments can easily be incorporated into a musical texture that otherwise consists of electrically produced sounds.

In relation to music genres, I focus on those dance music genres that emerged in the 1980s, such as house, techno, garage and rave (Thornton, 1996; Reynolds, 1998; Gilbert and Pearson, 1999), although there are important links to earlier dance music cultures such as disco that need to be acknowledged (see Straw, 2002, for example, for a discussion of the important role played by 12-inch vinyl in disco). Whereas these genres were and are often represented (by popular music scholars as well as participants) as belonging to particular communities or scenes (Straw, 1991), subsequent developments have questioned this logic through a fragmentation of dance music genres into various micro-genres. Aesthetically, the past decades have witnessed constant genre shifts that have deconstructed and reconstructed established genres such as house and techno, while introducing new sounds and styles. Organizationally, electronic music production is characterized by thousands of small production nodes, and the role of large conglomerates in this production process is a marginal one. Hesmondhalgh (1996) is correct in arguing against an optimistic view of flexible specialization as based on mutual trust and cooperation, but the hierarchical types of partnership he identifies between large and small music industry firms are simply not that important within electronic music networks. Technologically, electronic music production is substantially shaped by the rise of the internet as an important mechanism of distribution (Jones, 2002). This chapter largely concentrates – due to the period of data collection (2006–7) – on the current state of electronic music, but incorporates these more historical reflections on the development of electronic music in order to emphasize the network-specific path dependencies of music production that are irreducible to the urban environment (including the policy attempts to institutionalize creative clusters) in which this production takes place.

Creative clusters as policy imaginaries

Not only in London and Berlin, but also in cities as diverse as Amsterdam (Bertolini and Salet, 2003), Singapore (Lee, 2004) or Glasgow (Turok, 2003) – to name but a few – the debate on the creative industries has had a profound effect on policy discourses concerning the economic exploitation of urban cultures.

Based on the assumption that creativity *will* become increasingly important in the emergent knowledge society, policy makers have constructed an imaginary in which the creative industries are supposed to improve the competitive position of urban post-industrial economies. In Berlin, these discourses are often produced by actors associated with Projekt Zukunft (Project Future), which is a local government initiative 'devoted to structural change which lays the foundations for an information and knowledge society' (Projekt Zukunft, 2006). In London, most of the policy research has been undertaken by the city-wide governing body, the Greater London Authority (GLA) and it is guided by the same assumption. According to its report *Creativity: London's Core Business*, for example, we are dealing here with 'a fundamental transformation of London's economy' (2002: 5).

Many narrative themes are enrolled in this attempt discursively to position the creative industries at the forefront of economic development: from repeatedly stating that the number of so-called creatives is likely to grow (GLA, 2004a; LDA, 2006; Mundelius, 2006; Stadtforum Berlin, 2006), to celebrating the entrepreneurial character of cultural workers (GLA, 2002, 32; Projekt Zukunft, 2006: 19) or by emphasizing that the goals of social inclusion and cultural diversity are fully compatible with and achievable through economic development (GLA, 2003; Stadtforum, 2006). The debate on creative clusters constitutes one of these narratives. Building on Jones (1997; 1999) and Jessop's strategic-relational approach (1990; 1996; 2001), Brenner (2004) has suggested that state institutions are endowed with distinctive spatial selectivities, which leads them to privilege certain spaces at the expense of others and to channel socio-economic activities into these privileged areas. Following Brenner's argument, it becomes clear that the policy imaginary of creative clusters is strongly driven by this spatio-economic logic. Thus, in both London and Berlin creative clusters are seen as 'the focus for investment and support' (LDA, 2006: 30), a 'rationale for investment' (Creative London, 2003: 33), a way of 'strengthening the strengths' (Mundelius, 2006: 201 fn. 185) or as 'strategic spaces' on which to concentrate attention and resources (Stadtforum, 2006: 10). Based on the belief that the creative industries 'have a well-established reputation for playing a multi-faceted role in the regeneration of economies and environments and in supporting strategies for social inclusion' (GLA, 2004a: 139), the Creative London program has identified ten clusters in London with a high concentration of creative industries. These 'Creative Hubs' tend to be administered by borough-level and publicly funded economic development agencies who work together with a variety of private actors and whose focus is to offer services and facilities for cultural entrepreneurs. Similarly, in Berlin the support of urban clusters (*stadträumlicher Cluster*) is explicitly identified as a central field of action (Projekt Zukunft, 2005). Although it is acknowledged that more spatial clusters would need to be investigated, so far most attention has been paid to the development of the Osthafen, an area on either side of the Spree River in the eastern part of Berlin. A public–private partnership called Media Spree has been founded – involving a cooperation between real-estate investors, the Berlin Senate, adjacent boroughs and the Berlin Chamber of Commerce and Industry (IHK Berlin) – in order to channel public subsidies into branding strategies and infrastructural investments (for a more extensive analysis of these policies, see Van Heur, 2009a).

From clusters to networks and back again

In these and other policy imaginaries, creative clusters are seen to be relevant for all forms of cultural production. This assumption ignores, however, the much more complex spatial dynamics of many, if not most, forms of creativity, which in turn leads to flawed analyses of cultural practices. In this section, I point to the limits of analyses that see creative clusters everywhere by introducing empirical data on electronic music in London and Berlin and by highlighting the co-constitution of clusters and networks. In describing this mutual interdependence, I emphasize multiple values and determinations beyond the economic.

First of all, however, it needs to be acknowledged that it is partially under-standable as to why policy makers promote a localist cluster perspective, since a superficial observation of cultural production indeed seems to confirm the argu-ment that creativity emerges from local interaction and concentration. In the case of music, both London and Berlin contain a number of spatial agglomerations of electronic music production. In London, one can find clear concentrations throughout the city center: from Soho, Clerkenwell and Liverpool Street in the central area to Hoxton Square and Shoreditch in the eastern part, North Kensington in the western part and the South Bank and Bermondsey in the south. In Berlin, spatial concentrations can be found in Mitte, Prenzlauer Berg, Friedrichshain and Kreuzberg, all districts in the eastern part of the city. A somewhat more detailed analysis, however, reveals the much more complex socio-spatial dynamics behind the emergence of these clusters. Drawing on the theoretical literature on clusters, it becomes possible to identify two main dimensions of clusters: vertical and hori-zontal linkages; and knowledge and learning. Here, I will use these dimensions to analyze the data on electronic music and to show the often limited extent to which changes *in* the apparent cluster are the result *of* the cluster (for a more extensive discussion, see Van Heur 2009b).

Vertical and horizontal linkages

A central assumption of cluster theory is that – following Richardson (1972) – clusters are constituted by vertical as well as horizontal linkages between firms or actors. The vertical dimension consists of nodes that are functionally dissim-ilar, but that carry out complementary activities – a situation often described as a production system of input/output relations. The development of a cluster will lead to a process of differentiation in which suppliers emerge who cater to one particular process within this production system. The relations between these various nodes tend to be based more on cooperation and less on competition, since they are not competing for the same customers. On the contrary, it is the interac-tion between these nodes that leads to an efficient and economically effective cluster. The horizontal dimension of clusters consists of nodes undertaking similar activities and the relation between these nodes is therefore based on competition, since the success of one actor or firm will be at the expense of others. Nodes, therefore, are involved in a continuous monitoring and observing of other horizontally positioned nodes, since their own survival depends on being one step ahead of the competition. This tends to create a situation in which actors will copy successful competitors while adding some elements of their own – as a result, a self-reinforcing process of variation is set in motion. Central to this explanation is the argument that small firms are particularly dependent on these cluster-based linkages, since they are seen to rely more than large firms on their immediate urban surroundings. In his otherwise excellent work on Berlin, for example, Krätke argues that small-scale production tends to take place within local spaces and that large-scale firms connect these local spaces to a global cultural economy. As Krätke puts it, 'the global players of the culture industry network locally with the small specialized producers and service providers and simultaneously form a global network of its branches and affiliated companies, through which the urban centers of cultural production are connected with each other globally' (2002: 95). This rather dichotomous view leads him to assume – as does most cluster literature – that the 'creative atmosphere' as well as actual cultural products are the effect of clusters: 'The places of production of lifestyle-images are urban clusters of cultural production, which form above all in large cities' (ibid., 2002: 243). What this ignores, however, and what the data on music production in Berlin and London show, is the extent to which small-scale producers have become global players themselves.

Obviously, cluster-based vertical and horizontal linkages do exist in the case of electronic music in London and Berlin. Functionally dissimilar nodes that cooperate in order to create a functioning production system can be seen everywhere and the organizational fragmentation of electronic music into many small nodes can be understood as a process of differentiation in which each small node caters to a particular process within the overall production system. An obvious illustration would be the record label–artist relationship. Many labels in electronic music are very small, but in most cases labels represent not only the label owner, but also a handful of other artists who are regularly based in the same city. In London, to name but one example, Accidental Records started as a platform for Matthew Herbert, but after a while also released albums by other artists, such as Mara Carlyle, Mugison or Max De Wardener – all of whom were or are based in London for a certain period of time. In Berlin, Bpitch Control was founded by Ellen Alien and releases her work, but is also very much associated with Berlin as an electronic music city due to releases of other Berlin-based artists such as Apparat, Kiki, Modeselektor and Sascha Funke. Horizontal linkages – connecting similar and therefore (according to cluster theory) competitive nodes – are clearly visible as well. Venues and event organizers, for example, tend to be very much aware of the activities of other venues and organizers in the same city and one can therefore observe these nodes copying successful competitors, while adding elements of their own – thereby setting in motion a self-reinforcing process of variation and innovation. This partly provides an explanation and urban grounding of the ways in which particular music genres develop – something that has hardly been addressed by current analyses of popular music (e.g. Negus, 1999; Borthwick and Moy, 2004). For example, the dubstep genre in London started out as a label-oriented practice (including labels such as Tempa, Horsepower Productions, Big Apple) and was largely a virtual phenomenon in the sense that its main visibility was online through websites and forum discussions. One of the first club nights – titled Forward>> – was held in the Velvet Room in Soho (although it later moved to Plastic People and in 2006 to The End) and offered a space in which producers could present their sounds to a live audience. This was 'copied' by the DMZ club night at the Mass in Brixton and increasingly (although often mixed with other genres) by club nights at a host of other venues, such as Bar Rumba, Rhythm Factory, The Telegraph, and Electrowerkz (Clark, 2006; O'Connell, 2006).

At the same time, a second look at the various electronic music nodes must lead to at least a partial questioning of these cluster-theoretical premises. The main problem with the argument that clusters are shaped by cluster-based vertical and horizontal linkages is not that it is wrong to attribute emergent dynamics to clusters, but that this obscures the important role played by creative networks in co-structuring creative clusters. To recognize this, one needs to be sensitized to the aesthetic and cultural dimensions of electronic music production. The relation between musicians, for example, can be analyzed as horizontal linkages, with each artist standing in a competitive relation to other artists, but – even when accepting this questionable assumption of generalized competition – this tendentially ignores the relative status of each artist. Thus, a club night will often involve a DJ-set or live performance by one or more artists *not* based in the city – but visiting the city as part of a tour – backed up by a majority of DJs and artists who are based in the city. Quantitatively, the 'locals' outnumber the touring artists, but qualitatively the latter will play a much more central role within the club night than will the former. Similarly, the vertical linkages established between record labels and artists can be cluster-based (as discussed above), but often they are not. As if to illustrate this point, the Berlin label City Centre Offices provides a map on its website on which the artists' places of residence are marked. Besides Berlin (16 artists), the label releases music from Dresden (1), Munich (1), Heidelberg (3), Frankfurt (2), Dortmund (1), Lüneburg (1), Hamburg (2) as well as St Petersburg (1), Helsinki (1), Uppsala (1), Gothenburg (1), Copenhagen (4), Barcelona (1), Madrid (1), London (1), Northampton (1), Nottingham (2), Sheffield (1), Manchester (3), Woodstock (1), Toronto (1), Gainesville (1), Detroit (1), Los Angeles (3) and Mexico City (1) (all data from 3 September 2008). In other words, 16 artists from Berlin and 37 from the rest of the world. Naturally, the quantity of releases on medium-sized and large labels will mean that they are likely to include a sizable number of artists from outside Berlin, but there is no reason to assume that small-scale labels – which are part of the creative atmosphere that is so often celebrated in policy-oriented literature – are in any sense more dependent on cluster-based linkages.

Knowledge and learning

This brings us to a second characteristic of clusters and one that is increasingly theorized within cluster theory: clusters as the spatial prerequisite for the creation of knowledge and learning. The focus on the importance of knowledge – and above all tacit knowledge – in the emergence and reproduction of clusters has been put forward as part of a purported shift towards a knowledge economy and the development and increasing ubiquity of communications technologies. The basic argument runs as follows. In a condition of globality, in which everyone can have access to codified knowledge, the production of new and innovative products or processes is fundamentally dependent on tacit knowledge (Maskell and Malmberg, 1999). In contrast to codified knowledge, however, tacit knowledge doesn't travel as easily, since it cannot be expressed in signs (such as images or text), but is experiential and only partly conscious. As Gertler (2003: 79) has pointed out, there are two other closely related elements to this argument. One is that this local nature of tacit knowledge makes it 'spatially sticky' (also see Markusen, 1996), since the exchange of this knowledge between actors or firms can only take place if they share a common social context, which is largely locally defined. Related to this is the second element, which is the importance of '*socially organized* learning processes', since innovation is now increasingly based on the interactions and knowledge flows between firms and other institutions, such as research organizations or public agencies. Following this line of thought, other commentators have pointed out that particular activities within the creative industries are constituted by tacit knowledge, learning-by-doing and local skills (Crewe and Beaverstock, 1998; Leadbeater and Oakley, 1999; O'Connor, 2004; Raffo *et al.*, 2000). Although in the literature not a great deal of emphasis is put on explaining what constitutes knowledge and learning, the transmission of technical skills and the role of entrepreneurial knowledge are always understood as being of central importance (Wolfe and Gertler, 2004).

To assume, however, that the transmission of technical skills is cluster-based in the case of music production is highly questionable. This assumption seems to be a 'left-over' of the founding research on high technology clusters in which technical skills are communicated between research institutes and private firms, but is not directly applicable to creative clusters in which technical skills have a crafts-oriented quality and are much less part of complex organizational structures. Although this does not deny the local embeddedness of these technical skills, it does grant a relatively high level of individual autonomy and flexibility to actors that is incomparable to other industries (but typical for the cultural industries – see Hesmondhalgh, 2002; Ryan, 1992). One example should suffice. In recent years, the debates on open source culture – guided by the idea that cultural content should be freely available and accessible to all – have also left their mark on music production and quite a number of musicians are now involved in programming and using open source audio software. The basic assumption underlying open source software is that the source code is publicly available under a license that permits users to freely use, modify and distribute the software. Thus, a real-time programming environment such as PD (Pure Data) – based on the Max programming language and used for the creation of digital audio as well as audio/video projects – was originally developed by Miler Puckette, but is conceived as free software and is constantly extended by many other artists and developers. Similar to Linux (Mackenzie, 2005), therefore, PD as a software object concentrates social action and acts as a matter of concern (Latour, 2004). Clearly, this concentration of social action – involving both codified as well as tacit knowledge (through the embodied act of programming and making music) as well as actors in differing social contexts across the world – is by no means cluster-based, but instead part of global networks. Also, these networks challenge the economic discourses particularly prevalent within creative industries policy circles by constantly highlighting the importance of PD as a community effort and as collaborative work. Workshops – focused on learning, collaboration and experimentation – 'translate' these concerns to a local level and are regularly organized around the world. In London, this has been undertaken by networks such as GOTO10 and OpenLab. In Berlin, workshops on PD have been organized by the research centre 'xxxxx' (for an insider description of these scenes, see Mattin, 2005). These workshops do give a local (and potentially cluster-based) grounding to music production, but it would quite clearly be wrong to conclude from this that knowledge and learning are emergent from clusters.

The production of entrepreneurial knowledge has been put forward as another important quality of clusters. Within the cluster literature the notion of entrepreneurialism tends to refer to an awareness of market conditions and opportunities, personal responsibility, risk taking and a drive to achieve and grow. The background assumption of much of this literature is usually a meritocratic view of society in which achievement and the cultivation of social capital will pay – irrespective of structural inequalities (Somers, 2005). Despite the obvious need to criticize such a notion of entrepreneurialism, in the research on London and Berlin it became quite clear that at least some actors in the music networks had adopted these neoliberal discourses as their own. In that respect, the discussed music networks do not seem to differ from recent developments in other creative industries sectors (e.g. Manske, 2006; Neff *et al.*, 2005; McRobbie, 2002). For example, during an interview with Heiko Laux, the owner of the Kanzleramt record label in Berlin, he acknowledged that making a living by selling records had become increasingly difficult, due to free downloading and the enormous increase in record labels, but the solution he offered to overcome this problem was almost prototypically entrepreneurial:

> I have quasi redefined the function of the label. This resulted in new rules. The brand is, in principle, indestructible, even when the hard copies would cease to exist, the name is still there, and I would still try to bundle the music and get to the point. Thus, the transition to this digital business can in principle be mastered. [...] One simply has to accept the permanently changing rules of the game and adapt. One has to be capable of adapting.
>
> (Interview, 8 May 2007)

Although such a strong statement was rare, many actors adopted a comparable position, with statements such as 'When you love what you do, you find a way!'

(Possible Music distribution, interview, 23 March 2007) not being out of the ordinary.

Nevertheless, it remains important to emphasize the differing positionalities (Sheppard, 2002) within the networks of electronic music production, since many actors are not usefully understood as entrepreneurs at all. Theoretically, the argument that all actors are entrepreneurial risk takers and innovators amounts to voluntarism and to a 'semantic of total mobilization' in which cultural workers are always seen to aim for one step beyond the status quo (Bröckling, 2007: 117). The only way to avoid this analytical bias is to emphasize the contextual and habitual dimensions of social life in which innovation is an incremental process. Empirically, such a blanket statement is clearly wrong and it downplays virtually all those social and aesthetic dimensions of music that are irreducible to economic innovation, but which often constitute the very reasons for participating in music production in the first place. One obvious – but underdiscussed – phenomenon that questions the supposed centrality of entrepreneurial knowledge is the important role played by unpaid labor in shaping music production. For one thing, many actors combine their unpaid labor within music networks with paid labor in other fields. One of the main biographical solutions to the structural problem of underpayment in the creative industries is – besides family support, loans and unemployment benefits – working in sectors that are more secure and stable. Thus, Sam Shackleton – of the Skull Disco record label in London – works as a primary school teacher for 25 hours per week (interview, 12 December 2006). Eric Namour from [no.signal] – which organizes experimental music events – mentions that this project probably earned around £1,500 in 2006, but simultaneously emphasized that all income – if any – is reinvested in the production costs for forthcoming events. His main income is derived from a position within online music distribution (interview, 21 September 2006). Also, and parallel to the cross-subsidizing of activities, an important role is played by unpaid labor *within* each organization. On the one hand, this includes the acceptance that the work involves more hours than can be paid. On the other hand, this refers to the inclusion of unpaid labor by using interns or volunteers. Thus, the distributor Possible Music employs the owner, one part-time seller and two trainees, who tend to work from 11 a.m. to 6 p.m. (interview, 23 March 2007). The *Sound Projector* magazine in London operates as a non-profit undertaking and is supported by a handful of contributing volunteer writers (interview, 13 December 2006). Similarly, the Zur Möbelfabrik in Berlin is run as a members' club (*Verein*) and, although it does enable the main coordinator to make a living from the activities in the venue, most other workers – such as the bartenders – work as volunteers.

To an extent, these examples can easily be interpreted as reflecting a neoliberalization of labor conditions: as a result of increased competition, self-responsible actors accept low or no payment in the hope that they eventually will find a slightly more stable position (Ursell, 2000; Neff et al., 2005; Miège, 1989; Menger, 2006). This critical explanation, however, cannot explain everything. It tends to ignore, above all, that the high amount of free labor investment constitutes not only a mechanism through which workers can be and are exploited, but also labor that is 'willingly given' (Terranova, 2004: 94), since it offers actors the opportunity to participate in networks of pleasure, affect and collaboration. Not being paid (enough), in other words, also reflects a fundamental refusal on the part of actors to approach creative labor as a 'normal' job characterized by the sale of labor power. The notion of the creative cluster, in this view, is best understood as an abstraction: it abstracts from the multitude and heterogeneity of music practices and constructs a policy imaginary populated by economic entrepreneurs. Although I have no problems with abstraction as such, this misrecognizes the dominant tendencies in cultural production (economic as well as social, political, cultural, ethical, aesthetic etc.) and too often sees economic action where there is none.

Beyond economism: on the limits of policy

This brief chapter has tried to re-examine some of the popular claims concerning the supposed centrality of creative clusters in contemporary cultural production. After a description of the role of clusters as policy imaginaries in the cities of London and Berlin, the main section analyzed the actual dynamics of music production in these two cities and showed the limited extent to which this form of creativity emerges from clusters. Instead, I argued, electronic music production is best understood as shaped by the intertwinement of spatially concentrated clusters and spatially distributed creative networks.

The analysis, in other words, directs attention to the often limited explanatory value of the 'economic imaginaries' (Jessop, 2004) of policy makers. Three main consequences follow from this. First of all, it opens up a conceptual space in which one can highlight the limits of economic analyses of cultural and aesthetic practices. This space can be used both to develop a critique of the hype around the creative industries and to improve the explanatory relevance of policies. Second, it allows us to acknowledge the often limited impact of creative industries policies on cultural producers. The years since 2000, in particular, have seen a dramatic increase in policy institutions and governance networks dedicated to the creative industries, but the 'lack of fit' between their claims and the realities on the ground often makes it difficult to implement these policy proposals in practice. The complex self-organizational and trans-scalar dynamics of cultural production further exacerbate this situation. Third, however, the role of policies in developing new economic imaginaries – that select and privilege certain activities at the expense of others – does sensitize us to the importance of future-oriented projections in regulating social change. A critique of creative industries policies, therefore, cannot remain satisfied with a defensive rejection of these economic imaginaries, but will also have to include a vocabulary that can aid us in the development of new social imaginaries beyond economism.

9　Remediating vernacular creativity[1]

Photography and cultural citizenship in the Flickr photo-sharing network

Jean Burgess

Introduction

I originally began to use the term 'vernacular creativity' in order to draw attention to the material and everyday characteristics of the activities that sit behind the growth of 'user-created content', in contrast to the high levels of technological determinism that often underpin the more hyperbolic discourses around 'Web 2.0' and the 'democratisation of technologies' (Burgess, 2006; 2007). Vernacular creativity includes a range of mundane but intensely social forms and practices–from scrapbooking to family photography and the storytelling that forms part of casual chat. These are forms of creativity that predate digital culture, but that are shaping and being reshaped by the publicness of online social networks (like YouTube and Flickr) built around user-created content. While the growth of user-created content online is often represented as a radical social transformation, it could also be understood as the increased mediation and public visibility of creative practices that have long traditions in ordinary life.

Photography is a particularly rich site for the exploration of remediated vernacular creativity. Personal, domestic and amateur photography saw a dramatic period of mass popularisation alongside the development of the handheld camera, new business models and the Kodak brand at the turn of the last century (Jenkins, 1975; Slater, 1991); and at the turn of the present century, user-created content is flourishing alongside the development of technologies and business models oriented toward personal media production. Snapshot photography is deeply embedded in everyday domestic, leisure and familial life; in Flickr, a poster child for 'Web 2.0', personal photography converges with participation in online social networks to produce new forms of public culture.

Focusing on the remediation of personal photography in the Flickr photo-sharing website, in this chapter I treat vernacular creativity as a field of cultural practice; one that does not operate inside the institutions or cultural value systems of high culture or the commercial popular media, and yet draws on and is periodically appropriated by these other systems in dynamic and productive ways. Because of its porosity to commercial culture and art practice, this conceptual model of 'vernacular creativity' implies a historicised account of 'ordinary' or everyday creative practice that accounts for both continuity and change and avoids creating a nostalgic desire for the recuperation of an authentic folk culture.

Moving beyond individual creative practice, the chapter concludes by considering the unintended consequences of vernacular creativity practiced in online social networks: in particular, the idea of cultural citizenship. Before moving on to a detailed discussion of Flickr, however, it will be useful to work through some of the conceptual terrain that brings together the idea of vernacular creativity with the practice of photography.

Vernacular creativity and personal photography

The category of 'vernacular photography' has seen a recent increase in both popular and curatorial interest. The cultural work that this term does in increasing the visibility and legitimacy of particular forms of vernacular photographic practice resonates on several levels with the term 'vernacular creativity'. It is also substantively relevant because of the way that attempts to pin down vernacular photography reveal the contingency of categories of cultural practice *outside* the symbolic boundaries of official art worlds.

Geoffrey Batchen defines vernacular photography in the following way:

> The term 'vernacular' literally means the ordinary and ubiquitous but it also refers to qualities specific to particular regions or cultures. Its attachment to the word 'photography' allows historians like myself to argue for the need to devise a way of representing photography's history that can incorporate all its many manifestations and functions. A vernacular history of photography will have to be able to deal with the kind of hybrid objects I describe above, but also with, for example, photographies from outside Europe and the U.S. It may mean having to adopt non-traditional voices and narrative structures. It will certainly mean abandoning art history's evaluation system (based on masterpieces and masters, originality and innovation, and so on). In short, the term 'vernacular photography' is intended as a provocation and a challenge.
>
> (2002)

Based on this range of uses, then, the word 'vernacular' captures several important qualities of everyday creative practice. As with vernacular photography, speech or architecture, vernacular creativity is *ordinary*, as in non-elite and grounded in the materiality and experience of everyday life. But there is also a dual meaning of 'ordinariness' in Batchen's definition of vernacular photography that I mean quite deliberately to capture. Vernacular creativity, in being ordinary, is not elite or institutionalised; nor is it extraordinary or spectacular, but rather it is identified on the basis of its *commonness*. On the other hand, just as particular vernacular verbal expressions are indigenous to their temporal, social and geographic contexts, particular forms of vernacular creativity are grounded in contextual *specificity*. Indeed, as Anna McCarthy (2006) notes, from Raymond Williams' famous (1958a) statement that 'culture is ordinary' to Richard Hoggart's (1957) *The Uses of Literacy* and beyond, the uses of 'the ordinary' in cultural studies have always captured this apparently contradictory duality. McCarthy (2006: 34) writes that 'ordinariness' has designated both 'the embodiment of concreteness' in describing 'the sediment of practices that make up everyday life on the small scale of lived experience' and the very large category of things that were not extraordinary or special (literally, 'out of the ordinary'), and therefore not rare or scarce.

Precisely because of the rich meanings of both 'everyday life' and the 'ordinary' as I use them here, it is important to establish that I am not using the term 'vernacular' in order to create an aura of authenticity or purity around everyday creative practices. Based on cultural studies scholarship, I understand vernacular creativity to be bound up with, not separate from, popular consumption and engagement with popular culture, because popular culture too is profoundly embedded and made use of in everyday life. Additionally, while the domain of vernacular creativity is everyday life and not the institutions of 'official culture' or the production end of the creative industries, at the same time it often operates with reference to the values, aesthetics and techniques of established creative professions and art worlds (Howard, 2005).

In his call for a 'vernacular theory of photography' Batchen (2001: 59) discusses the relations between vernacular photography and 'proper photography'. In doing so he sets up the vernacular as official art history's 'other', describing vernacular photography as 'the absent presence that determines its medium's historical and physical identity; it is that thing that decides what proper photography is not'. Batchen's argument that vernacular photography has to be theorised on its own terms, and not as a kind of 'lack' as viewed from the perspective of 'official' photography, is tremendously useful. But, as Fine's (2003; 2004) work on the ideology of authenticity in the construction of symbolic boundaries around 'self-taught art' reminds us, it is important that attempts to imagine, describe or celebrate the forms and practices of 'vernacular creativity' do not at the same time contribute to the drawing up of boundaries which result in its symbolic exclusion from the domain of 'real' creativity, and therefore the perpetuation of the existing order of cultural legitimacy.

Further, such distinctions are unproductively artificial: as I discuss later, the boundaries between vernacular creativity and art, and between vernacular creativity and commerce, are continually being blurred. Nowhere is this more true than in convergence culture online, where the emerging big media behemoths like Yahoo! and Google actually structure their business models around the provision of platforms for ordinary people's creative practice and collaboration; vernacular creativity is not 'outside' the market; it is driving and being driven by it. Rather than seeing this development as the capture of authentic, resistive vernacular creativity by corporate interests, it is more productive to see this convergence of 'market' and 'non-market' cultures as part of the conditions within which the contemporary remediation of vernacular creativity takes place. Vernacular creativity drives the development and produces much of the value of the online creative networks that form part of the 'Web 2.0' model of online participatory culture – as exemplified by Flickr, the photo-sharing network.

The Flickr photo-sharing website

In 2006 the Yahoo! Finance website described Flickr, which it acquired a year or two after its launch in February 2004, as 'an online digital photo management service that allows users to store, search, sort, and share pictures from digital cameras or camera phones'. But although in business terms, of course, it is a service designed to bring users within reach of advertisers (albeit often indirectly in this case), Flickr was designed as a 'community' built around internal image sharing and social interaction from the beginning.

Prior to the launch of Flickr, developer Ludicorp's major enterprise was a massively multiplayer online roleplaying game (MMORPG) called Game Neverending. Foreshadowing some of the principal features of Linden Labs' enormously successful virtual world Second Life, Game Neverending (GNE) was primarily a virtual space that afforded open-ended forms of social interaction and object manipulation. It was the object-sharing, extensible architecture and live chat tools built for GNE that were later repurposed by the developers, eventually becoming the basis of Flickr. Because of this, at the most basic level, the sharing of personal photographic images as the basis of social interaction is explicitly written into the architecture of the service.

Flickr soon evolved into a fully fledged photo-sharing website, with a strong emphasis on camera phone images and the everyday. It was bought by Yahoo! in 2005 for $35 million, and it has remained something of a poster child for Web 2.0 and user-led content creation. In November 2008 Flickr claimed to host more than three billion photos.

At the level of its most basic functionality, Flickr is simply a 'photo-sharing' service – it is a website to which individual users can upload their personal photographs for others to view. However, as the analysis developed in this essay will demonstrate, Flickr's architecture has affordances that go far beyond the publishing and viewing of images, and extend to a number of levels of social and aesthetic engagement.

The participatory turn in web business models that the business and web design communities refer to as 'Web 2.0' (O'Reilly, 2005) is characterised by the convergence of social networks, online communities and 'consumer-created' creative content, a convergence which sometimes (especially in the UK) goes by the name 'social media'. Flickr is one of the first and most well-known examples of this trend towards convergence between online social networks and creative content distribution; it is in fact an emergent and collaborative three-way articulation of social networking with individual content creation and communities of practice – the emergent cultures that result from interaction between the members of the more abstract 'designed' community – in this case, the Flickr network's architecture and its affordances.

The following discussion of Flickr as a site in which vernacular creativity is remediated is grounded in over two years of participant observation, including exploration of and direct participation in the Flickr network, as well as research into the discursive framing of Flickr as a Web 2.0 'architecture of participation' in the blogosphere and other authoritative sites of web commentary. I joined the network myself in 2004, and began uploading my own images, adding contacts to my profile, leaving comments on other members' images, and responding to comments on my own images. My Flickr images were often cross-posted to my weblog, which was the hub of my online research practice as well as the locus of my participation in 'DIY' web culture. I joined several interest groups within

Flickr, but concentrated for research purposes on two: the Brisbanites group (for photographs taken in Brisbane) and the Brisbane Meetup group (used for organising offline photographic excursions and social occasions, and sharing photographs of these events). In order to develop an understanding of the ways in which people were integrating Flickr into their everyday lives and their photographic practice, from among the members of these two local groups I recruited seven participants with whom I conducted extended interviews, for the most part at their homes. I also undertook field trips to offline Flickr meet-ups in Brisbane, which were also attended by several of the participants I had interviewed.

The convergence of social networking and creative practice

Everyday snapshot photography is arguably the most persistent of all vernacular media forms, and all of its genres are represented in Flickr somewhere: baby and pet photographs; the documentation of family events; the visual record of holidays and the mundane details of everyday life. The point at which remediation begins is when these photographs are uploaded to an individual user's database and both become part of a shared public resource and facilitate the uploader's participation in a large social network. At the same time, those social connections are used to collaboratively construct, negotiate and learn visual aesthetics and techniques.

Flickr's database structures, and the way they are navigated via the user interface, can be seen as an 'architecture of participation' that shapes the culture of Flickr so that social connection is primary and formal aesthetics are secondary. In this context, 'content' is not a commodity that can be exchanged, but a means of social connection; at the same time, those social connections are used to collaborate, learn and teach photographic aesthetics and techniques. Flickr is at the same time a showcase for one's own work, an exhibition of everyone else's work, a social network and a learning community – or, simply a place to upload and share photos with family and friends.

In the following exchange about her particular mode of participation in Flickr, Louise, who is interested in the 'fun' and 'social' aspects of Flickr as much as the opportunities for aesthetic or technological development, weaves together online and offline social interaction, learning and play in her discussion of what makes Flickr an engaging space for her:

> It's fun along the way, the journey is the game, the journey is the fun. So I get a lot of fun out of it, I learn a lot, and it makes taking a photo more fun, because like my husband goes 'Oh, that's nice' but here – lots of people are seeing it and saying ... [for example] you've got a dirty lens! or 'that's great!' [...] And then you meet them socially, and you have a nice time, and you have a common thread of taking photographs. [...] So [Flickr] has great momentum, because there's a common theme.

Flickr is also a space where 'professional' photographic aesthetics, 'art' discourses and vernacular photography collide, compete and coexist. Partly because of the intensity with which Flickr promotes the 'personal' photo-sharing ethic of the network, most Flickr users are not 'professional' photographers, nor are they 'artists' in the sense of fully occupying those identities. However, the participants I interviewed for this project do represent themselves as creative practitioners, or even self-taught artists, and some harbour significant ambitions for their photographic work. Cyron framed his photography as a creative practice, even though he does not actually view himself as a 'creative person':

> Photography is probably the first time that I've ever felt a bit creative. My school life consisted of things like physics and chemistry and maths, and I couldn't draw a stick figure to save my life, I can't paint, I can't sing, and I don't feel the urge to do any of those typically creative things. […] It allows me to actually be creative, I can actually produce something, where the end result is that I can actually feel that I've achieved something creative, I've done something.

The reason photography 'allows' him to be creative is that it enables him to produce what is widely understood as 'creative content', rather than providing a vehicle for his 'innate' creativity. Cyron understands his continuing development as a 'creative' photographer as a technical, aesthetic and self-educative process that has enriched his everyday experience in particular ways:

> It has changed the way I look at the world because I wander around looking at things from different angles, from different perspectives, trying to – always bearing in mind what might make a good photo and, I mean, I don't profess to be a good photographer, I don't profess to be a bad photographer either, but whatever the case is, I find it very fulfilling for myself. And it's a good thing when other people do as well, when other people look at my work, which is what Flickr has been good for.

He goes on to imply that the acquisition of a digital camera, making it easier to capture and share images, combined with the social networking aspects of his engagement in Flickr, adds a layer of uses for personal photography beyond the recording of images, including the ability to publish and receive feedback on his work.

Of the seven participants I interviewed, those who were most invested in professional standards for photography, and who sought out opportunities to learn 'proper' photographic techniques, were least likely to represent the more 'private' aspects of their lives in public, beyond photographs of family members and personal occasions. Conversely, Melanie, who persistently frames herself as 'self-taught' and whose practice represents a refusal of institutionalised aesthetics, in her own words 'takes photos of everything' – hundreds of self-portraits and intimate snapshots of her family members, as well as photographs of flowers, food and her own artworks.

Melanie has her own online store where she sells her handmade jewellery, gift cards, textiles and tarot cards. Referring to the items for sale in the online store, I asked if she was self-taught in craft as well, to which Melanie responded 'Yep, of course'. In response to a further question about whether she had ever done any short courses, or even refers to books to learn how to do things, Melanie said she hadn't.

I just look at the pictures, and think 'Ooh, I like the look of that', and if I do it I will make it up myself, not look at the book and see how they did it. I'll see their idea but I won't copy their method, I'll make my own.

Melanie mentions the positive feedback she has received from people whom she sees as bona fide artists.

But it is very odd at first, that people keep saying 'Oh but you are an artist!' Like I say, I want to be an artist when I grow up, and [they say] 'you already are'.

All of the participants mentioned occasions when their images had become popular within Flickr by being favourited and viewed several times or even making it into the Explore pages (based on a high level of 'interestingness'). Beyond that, several of them had begun to develop aspirations beyond the intrinsic reward system of the Flickr network. For example, David had started to enter amateur photography competitions and found that the external rewards created additional motivation to pursue his interests.

Mr Magoo ICU, on the other hand, is very invested in the idea of art photography, but his practice takes place outside any formally constituted art world – he works literally underground. In his photographic practice he concentrates mainly on urban exploration photographs of Brisbane, both underground (tunnels, drains and sewers) and overground (abandoned and condemned buildings and construction sites), and is adamant about using lighting and in-camera techniques, rather than Photoshop, to create the surreal and atmospheric effects that characterise his images. Although he earns his living from IT, his photographic practice is far from a casual leisure pursuit, and he differentiates his work from personal or everyday photography as well as commercial 'stock' photography on the basis of his technical expertise, artistic sensibilities and professional aspirations, placing him squarely in the 'ProAm' (Leadbeater and Miller, 2004) category.

For most of the participants, the greater their investment and participation in the Flickr network had become over time, the greater their interest in producing 'good' images became also. Throughout the participants' discussion of their photographic practice and its development is the underlying assumption of an ideal progression from everyday documentary photography – photographing the family, family events, holidays, or just 'photos of everything' – to 'better' or 'more advanced' photography. However, this does nothing to weaken the core affordances of Flickr as a social network – however deeply invested the participants might become in photographic aesthetics, techniques and technologies, it is photo *sharing* that constitutes participation in Flickr, over time creating hybrid networks of fellow creative practitioners, friends, family and acquaintances.

'Share your photos. Watch the world': Flickr and cultural citizenship

The organising discourses around Flickr focus on the interplay of the personal and the everyday with broader cultural communication and creative practice

– an interplay captured nicely by the current slogan 'Share your photos. Watch the world'. Quite early in the evolution of the Flickr network, industry expert Tom Coates captured the range of emergent cultural practices that coexist on the network in the following description of Flickr as an example of social media:

> On Flickr many people upload photos from their cameras and mobile phones not just to put them on the Internet, but as a form of presence that shows their friends what they're up to and where in the world they are. Their content is a social glue. Meanwhile, other users are busy competing with each other, getting support and advice from other users, or are collecting photos, tagging photos or using them in new creative ways due to the benefits of Creative Commons licenses.
>
> (Coates, 2006)

As the interview material discussed above shows, participation that begins with casually storing and sharing family photos with an existing personal network can and does evolve into a more ambitious engagement with photography as a craft and a form of creative practice; deeper engagement with photographic techniques and practices leads to more intensive social interaction and investment in the idea of Flickr as a community, with the possibility of both intensely local and globally distributed interaction.

It is important to point out that the practice of cultural citizenship in Flickr is constituted not only online, but also through the articulation of the 'online' social network with everyday, local experience. One of the most important features of Flickr is the ability to create 'groups' – communities of interest and practice – within the network. There is one such group, Brisbanites, for uploading and discussing images of Brisbane. As well as hosting photos of everyday life, tourist images, and photographs of urban decay, recently the group became the locus of vernacular history when an Italian user known on the network as 'Pizzodesevo', now resettled in Italy but who had been resident in Australia in the 1950s and 1960s, began posting scans of slides taken at that time to the Brisbanites group.[2] A number of group members showed interest in the photographs by leaving comments that ranged from expressing appreciation, to offering technical advice about scanning, to discussion of the locations of the photographs and how much they had changed in the past 46 years. The connections made between users as part of this discussion resulted in one Brisbane-based member of the Brisbanite group spontaneously creating a kind of game around the images: he began going out specifically to capture images of the same locations as in the old slides, and uploading them to his own Flickr photostream. 'Pizzodesevo' then combined some of these new images side by side with the old ones in a series of diptychs that reveal the often dramatic changes to the Brisbane cityscape, which in turn led to more discussion about the ways in which the city has changed, blended with nostalgia for a past that many of the discussants had never encountered themselves.

In another example, Louise, one of the participants in this study, started the 'Themed Flickr Photos' group,[3] where users collaborate to establish a subject that might be found in almost any city in the world, go out to capture images of that subject in their own towns, and then upload them to the group photo pool. Past 'themes' have included McDonald's restaurants, traffic and shoe stores. Louise told me that she had the idea after coming across a photo from Canada of a place called the Tongue & Groove. Remembering that there was a Brisbane restaurant with the same name, she drove to West End and took a photo of it, uploaded it and let the Canadian photographer know by leaving a comment on his original photo. The Canadian photographer suggested that she expand the idea into a group where common features of cities on opposite sides of the globe could be photographed and shared in the group pool.

Members of the Brisbanites group have also begun organising regular offline meet-ups – opportunities for socialising combined with photographic expeditions in the city, suburbs or surrounds. All but one of the participants I interviewed had attended at least one of these occasions. The ongoing participation in meet-ups has several effects: the cultural practice of 'belonging' in the city, especially as the photographs of the meet-up and other Brisbanites' photos are circulated as vernacular representations of 'Brisbaneness' in the cultural public sphere; intensified and more meaningful everyday creative practice via the collaborative photographic excursions; and an intensification of the 'community of practice' (via comparing uploaded images of the meet-up, as well as members giving each other technical and aesthetic feedback and advice); recursively enriching both online and offline social and aesthetic engagement. As an example, one of these meet-ups took place as a ride upriver on the CityCat, beginning at the Regatta terminal and disembarking at the University of Queensland, which was at the time the construction site for a new and quite controversial 'Green Bridge' project.[4] As the group, made up of people with a range of ages, identities and occupations, wandered along the riverbank taking photographs and talking, the conversation flowed seamlessly between a wide range of topics: comparisons of cameras, advice about technical settings for particular shots, the beauty (or otherwise) of the surroundings, discussions about the Flickr network and, most interestingly, deliberation regarding the Green Bridge: What should it be called? Should it be there at all? Is it beautiful? Is it good for the environment? When will the 'golden hour' just before sunset start, so we can get a great shot of it? Upon returning home from the meet-up, most participants upload their 'best' shots from the event to the meet-up group's pool of images so that other group members can view and comment on them, often leading to a continuation of the discussions that occurred during the meet-up itself.

Such participation takes the form of what Habermas (1996) terms 'episodic publics' – the ephemeral everyday encounters in taverns or trains where citizens negotiate (or, in rationalist terms, 'deliberate') matters of shared concern; or 'occasional publics' – where groups of citizens gather for particular occasions (the rock concert, the public funeral). The mode of participation in Flickr that most obviously constitutes civic engagement is a convergence of 'offline' everyday life in a particular local context with 'online' participation in digital culture and with cultural and commodity consumption. That is, the representation of the local, the lived, the specific, flowing into the discussion or negotiation of discourse *around* those representations, and the participation in communities within the Flickr network that may reference or flow back into offline social lives.

These uses of Flickr hint at a more general possibility – that online networks organised around user-created content represent potential sites of an emergent cosmopolitan cultural citizenship. They constitute authentically everyday spaces in which individuals can self-mediate their identities and their perspectives on the world, engage with the self-representations of others, contribute to collective outcomes and encounter cultural or political difference.

Indeed, Flickr founder and former CEO Stewart Butterfield evoked a playful, everyday cosmopolitanism as an ideal, and at the same time connected everyday vernacular photography with the networked public sphere, in his announcement that Flickr aimed to be 'the eyes of the world':

> That can manifest itself as art, or using photos as a means of keeping in touch with friends and family, 'personal publishing' or intimate, small group sharing. It includes 'memory preservation' (the de facto understanding of what drives the photo industry), but it also includes the ephemera that keeps people related to each other: do you like my new haircut? should I buy these shoes? holy smokes – look what I saw on the way to work! It lets you know who's gone where with whom, what the vacation was like, how much the baby grew today, all as it's happening.
>
> And most dramatically, Flickr gives you a window into things that you might otherwise never see, from the perspective of people that you might otherwise never encounter.
>
> (Butterfield, 2006)

This is a very particular model of cosmopolitanism – bourgeois, individualised and relentlessly cheerful, coming as it does out of the West Coast technoculture. But Butterfield's normative description of what Flickr is *for* captures one aspect of the site's potential particularly well. That is, it is the sheer ubiquitous and ordinary nature of most everyday photography, and its embeddedness within social life, that creates the conditions for Flickr as a platform for vernacular creativity with all its unintended consequences – citizenship, commerce and art.

Notes

1   Some sections of this chapter have been adapted from material previously published in Jean Burgess, Marcus Foth and Helen Klaebe, 'Everyday Creativity as Civic Engagement: A Cultural Citizenship View of New Media,' Proceedings of the Communications Policy and Research Forum, University of Technology Sydney, September 2006. Available at http://eprints.qut.edu.au/archive/00005056/.
2   See, for example, www.flickr.com/photos/globetrotter1937/195304137.
3   The 'Themed Photos' group is at: www.flickr.com/groups/55001358@N00/.
4   The Green Bridge is a project of the Brisbane City Council. In an effort to reduce traffic congestion without negatively impacting the environment, the bridge provides a bus, pedestrian and cycle link across the Brisbane river between Dutton Park and University of Queensland, but is closed to cars and trucks.

Part III

Everyday spaces of creativity

10 Creativity, space and performance

Community gardening

David Crouch

Introduction

Creativity appeals because it is vital. In this chapter I examine ideas about the dynamics of creativity that embed it in everyday living: in things people do, how they get by, feel a sense of wonder and significance, and make or find becoming in their lives, personally and inter-subjectively. Creativity in everyday life is a dynamic through which people live. A particular consideration is the expressive character of creativity in everyday life: expression in materiality and in friendship, thinking and feeling. Throughout this discussion creativity is considered through the relational character of individuals and space.

I begin with a brief outline of the attention familiarly granted to the creativity associated with gardens. I note the shift in thinking about gardens to gardening, and how this occurs. Next I acknowledge privileged prevailing notions of creativity: the triad of making artwork, and the corporate and institutional making of creativity as exemplified in the creative city and creative industries; and a more recent debate on creativity amongst technologists and their materials. I argue for the distinctive character and importance of a different kind of creativity in everyday life; and along with the use of contemporary ideas concerning space, I outline particular philosophical positions for considering creativity in gardening, which exemplifies how creativity and space commingle and are mutually emergent. The kind of gardening highlighted here is popularly known as community gardening, or 'allotment holding' in the UK, and it opens up specific ways of thinking about creativity.

Creativity is a culturally loaded term that is familiarly elided with the arts, invention, and perhaps more recently, technology. In thinking of human life, I will explore the possibilities of more everyday human engagement in a complex world in terms of capacities, situations and processes of creativity.

Art certainly exemplifies some of the character of creativity. Considerations of artwork as creativity have tended to concentrate upon the object, the performance, the painting, the musical score and music as performed. Yet the creative process, that I argue is universal, happens to the artist in a much more complex and nuanced process of living (Crouch and Toogood, 1999). Hence I am interested in reflecting, later in this chapter, upon the character of process or performance in the longer process of living, through which, for example, a painting emerges.

Creativity is also familiarly cast in relation to the work of institutions, and particularly the commercial sector, for example in relation to the notion of the 'creative city' and cities of culture. In their wide-ranging discussion of economies of signs and space, Lash and Urry discuss possible ways in which 'the aesthetic reflexivity of subjects in the consumption of travel and the objects of cultural industries create a vast real economy', art galleries, bars, taxi drivers and brokers acting in the flows of diverse cultural producers and mediators (1994: 59). Yet this 'real economy' and its real creativity conceal the complexity of cultural economy. For the different components, spaces and events used are combined through the performative encounters and practices of the individual, perhaps as consumer or traveller. Through their performances, individuals work, select significance amongst a complexity of things, feelings, relations and actions, and affect the vitalism of the city. The apparent 'real economy' is thus worked and mediated through individuals' own multiple relations with these other resources of economy. Prevailing discussions of cultural economies have tended to underplay these more human components of contemporary cultural and geographical worlds in overemphasising the role of a more obvious cultural class.

For example, Sara Cohen (2007) shows the limitations of considering creativities only amongst institutional and business efforts to claim the cultural arena of Liverpool, City of (European) Culture, finding a wealth of creative pursuits among local music groups and individuals across the city. She contends that opportunities for more creativity might depend more on freedom, openness and a lack of institutionalisation outside the Allocated Core, the so-called 'Creative Quarter', yet she could find little concern among cultural policy workers to engage and listen to these musicians.

Academic thinking on creativity is also a focused upon the creative process, of academics themselves, of technologists and their work in industry. For instance, Nigel Thrift emphasises the exercise of creativity in highly self-conscious cultural economies, pointing to how the design of rooms in high-intensity technology centres encourages a 'creative sociability' in a mixture of cafes, informal meeting rooms and labs (2008: 44–5). Yet this creative sociability is rarely explored in 'everyday' worlds. Accordingly, I investigate forms of sociable creativity produced through gardening.

Gardens and the mundane

Gardens, or 'the garden', have been evocative of creativity in different ways (Crouch, 2009). The garden has been used to represent an ideal environment and culture, 'gardens of creation' signifying a pre-cultural, pre-human state in several religions. Gardens have also come to be associated with creativity in relation to the art of the landscape gardener who professionally designed features of land and material into different assemblages to produce a particular feeling or idea. Yet such productions are surrounded by subjective symbolism, for example through which a garden may signify a particular articulation of Englishness. More recently the garden is a creation of the middle class and its idea of domesticity, and gardens have reproduced particular ideas of nature (Franklin, 2001). There has been extensive use of gardens as objects of creativity, and to excite creativity across the arts. In all of these ways the garden extends from a creative assemblage of material artefacts to its creative production as an object.

In recent years increasing attention across social sciences and humanities has focused upon gardening as process, produced through practice and performance. This shift of emphasis is part of broader efforts to develop ways of thinking and articulating space and life together in a way that can attend to the nuances of being human in the world. My concern is with how as individuals we engage in the making of space in and across our different moments and how this engaging inflects significance in our lives. Though particular collective attributes of progressing life are frequently ignored, cultural economies are significantly built through mutual regard and care and its creativities (Leyshon, Lee and Williams, 2003), and Lee examines the practical shared knowledge among gardeners that produces continual change (2000). Accordingly, I explore the nuanced complexities of living, signified as micro-processes or micro-politics, with the implicit presumption about the mundane, that what matters lies elsewhere. This potential can be explored through gently excavating thinking and feeling, for example in new work on emotions and subjectivities.

Individuals do not stay ontologically fixed; things change in their everyday lives; their world views adjust, partly through small cumulative actions that can colour meanings, significances, attitudes. Thus, rather than through only an explicit or activist politics, a gentle cumulative politics of adjustments of thinking, feeling, mutual regard, regard for nature, and therefore adjustment of the character of practice can occur, and can be profound in its feeling and ideas. The nuanced and felt, and the coarser grain do not act in distinct camps, of course, but are mutually and multiply mingled in flows of activity. In this, there is much potential for progressive hope. An emphasis on 'positive' experience, feeling and thinking, suggesting 'mere celebration', is often treated warily and critically. My attention to the 'mundane' seeks not to essentialise or privilege, but to acknowledge; to let process breathe, to explain the entanglements in living, doing, thinking and feeling.

Here then, I focus on ideas of creativity through a particular kind of gardening: community gardening, allotment holding in the UK. In many ways this compares to cultivating a garden or yard 'at home'. However, community gardens are rented plots laid out across a site for a number of people to use. Thus there is a combination of need and desire for land, a distinctive feeling of ownership and sociality in community gardening. Indeed, community gardens emerged through political movements to acquire land for those who lacked opportunity to cultivate. This sociality may be welcome or unwelcome, but it can grant opportunities for intersubjectivity and social activities that may be different from working the garden at home. Moreover, a whole site offers a distinctive kind of space, a collection of people working in fairly close proximity, usually without strongly marked boundaries, that is more complex and nuanced than the average home garden, and individuals have an opportunity to learn from each other in these semi-public sites.

In the section that follows, I engage distinctly different conceptual ground from which to progress this discussion of creativity and the mundane. However, I argue that in each conceptual framework there are noteworthy points from which reflection and reasoning on this kind of creativity can make sense. The subsequent section develops this thinking in relation to ways in which creativity may be pursued, or may emerge, in relation with space. Here, the spaces of the particular 'mundane' creativity are opened up and unravelled, drawing upon the positioned conceptual orientations through attention to empirical evidence constituted in talk with individuals doing community gardening.

On creativity and cultural life

In their critical discussion on creativity Hallam and Ingold engage a sense of how creativity relates to people's lives:

> creativity is a process that living beings undergo as they make their ways through the world … this process is going on, all the time, in the circulation and fluxes of the materials that surround us and indeed of which we are made – of the earth we stand on, the water that allows it to bear fruit, the air we breathe.
>
> (2007: 2)

Creativity, they argue, emerges in cultural improvisation. This process is different from the familiar way of thinking of creativity as innovational. It also contrasts with the notion that what people do is organised through a fixed plane of routine procedures: there is, instead, opportunity and imaginative possibility. Our cultural life does not operate to a given script, nor is one available. Habitual and routine tasks necessitate adjustment and response, as the world and our particular circumstances and emotions are never the same. Accordingly, the vitality of things we do affects the possibility and character of creativity that may emerge.

In our lives we construct, handle, make sense, cope, respond and anticipate amid a complex collision of influences, unbidden occurrences and desires, only partly planned. Our way in life is 'continually altered and responds to the performance of others' (ibid.: 4). Creativity emerges through the experience or practice of doing. Temporally, it is not collapsed into an instant or a series of instants but embodies duration, multiply enfolding and unfolding. Creativity relates to life's emergent potential to shift and change as becoming, rather than as stasis or fixity.

The idea of performativity points to sharp breaks that mark significant change. Yet Hallam and Ingold's ideas highlight how flows of performativity can be non-acute, slow, gradual and accumulative. The work of Deleuze and Guattari, in contrast, gives a sense of urgency and acceleration to the notion of becoming. For them, 'becoming' is explicitly related to the countless possibilities of things, of life, emerging or erupting from an 'immanent surface' of energy (2004; Dewsbury and Thrift, 2004). Developing this argument for the acute intensity through which creativity emerges in living, Grosz (1999) elucidates how individuals live their lives in perpetual negotiation, operating in the tensions and tugs of 'holding on' and 'going further'. It seems that we unreflexively bear traces and desires of continuity amid change, and yet alongside and flowing awkwardly among these forces we sustain a desire for difference, adventure and novelty. These multiple 'tugs' are inherent to journeys of living and the dynamic of becoming happens in movements, in vital energies between different forms of intensity, low-level and hyper-intensity (DeLanda, 2004). Different kinds of energy-intensity among moments of encountering the world contain diverse potentialities, different kinds of significance and varied feelings, through which we return to familiar traits and open ourselves out to new experiences.

Deleuze and Guattari's notion of life as lived rhizomatically is significant here in thinking through how journeys in life happen; the spaces erupt along and among these laterals. Plants that have fleshy root-like structures work across ground laterally, rather than, at least primarily, in roots. This simile anticipates the multiple and lateral relationalities, rather than the deeply fixed and vertical structures, that they sought to apply to human life amongst other life, via networks rather than foundations. However, it is worth noting that plants with rhizomes usually possess roots too, albeit usually weaker than their rhizomes, through which, after emerging mobile on the ground, they may both 'hold on' as well as 'go further'.

These ideas resonate with the ways in which Ingold has discussed complexity of journeys in life through the notion of wandering (through a wandering mind as much as wandering feet) (2007). His lines meander, fold back, strike a different direction, and space tracked may summarily disappear to reappear later. Crucially, journeys have varying thickness and fleshiness, often uncertain, however habitually they may be tracked. Such journeys mix feelings, memory and the character of practice and performance. They also resound with Bergson's notion of duration, a mix of progress and heterogeneity, through which memory both holds on to the past but shifts (Bergson, 1911: 164). Bachelard, powerfully fleshing out his notion of time – he felt that instants of time became 'fleshed out and filled in later' (2000: 89) – similarly acknowledges the complexity of a heterogeneous time, diverse, vibrant and full of energy.

As a consequence, creativity can happen unexpectedly in the everyday, the ordinary and mundane, because each of these is open to accident, variety, disruption and change. Efforts can be made to sustain precision in practice but those efforts can only partially control. And we negotiate in ways that draw on both Hallam and Ingold's notion of creativity as everyday improvisation and the more insistent and dramatic notion of performance: negotiating life, between the dragging desire of holding on and an awkward, perhaps impatient desire to go further, elsewhere, to do differently.

Thus in the emergent and multiple character of duration in the more fleshy form of journeys creativity emerges. There is a tension between the notion of 'becoming' as bearing limitless and effusive energy, styled as contagion, and possibility without limit, to be anything (Thrift, 2008), and the idea that creativity is a more modest dynamic. Yet these tensions mirror the everyday experience of living: not only profound adjustment, but also subtle negotiations. Particular moments may thread across other areas of living from the particular site, act or moment through which they occurred.

While Deleuze and Guattari and others have pursued the more intense characteristics of creativity in the potential of becoming, the ever-present, immanent character of change, disruption and the unexpected, Hallam and Ingold's interest is in a slower character of performance but one no less significant. They point to sustaining life as a ground for creativity, as an ongoing component in the 'mundane' character of leisure activities in everyday living, its coping and adjustments on a more modest scale (2007). They point to the possibility of change, adjustment, recovery, and argue that 'there is no script for social and cultural life ... people ... improvise' (ibid.: 1). They draw out the generative rather than repetitive character of creativity, relational rather than novel. Thus creativity must be conceptualised in a world that is never the same from one moment to the other and in a way that pays full regard to so-called conventional creativity (ibid.: 2). Ingold exemplifies the creative move of individuals' lives through the example of a house, whose creativity is not fixed in the architect's realised design but is constituted through the living of those who dwell there. Creativity emerges through the persistently creative performance of tradition, rituals referred to changed circumstances, and 'can be truly liberating' (ibid.: 3).

Creativity and space

Doreen Massey (2005) argues that space is always contingently related in flows, energies and the liveliness of things; is therefore always 'in construction', rather than fixed and certain, let alone static. Thus what space 'is' is crucially rendered unstable, shifting matter and relations in process. It may be felt to be (more) constant and consistent, but that feeling is subjective.

I want to expand this notion of dynamic space by turning once again to the work of Deleuze and Guattari (2004), which has unwound the ways in which philosophies have tended to think about the vitality of living, the multiplicities of human and non-human influences in a world of much more than human subjectification and response. Their particular reasoning of space is of particular interest through its conception of the process-dynamics of space/life in spacing. What they mean by space is highly contingent, emergent in the cracks of everyday life, affected by a maelstrom of energies well beyond human limits. What interests them is the potential of space to be constantly open to change, becoming, rather than only the more settled. In these respects there is resonance with Massey. However, for them spacing works in the gaps, the relations and distances between and among things and things that happen and not only is referable to space as that inhabited metaphorically and materially by people but emphasises the character of things happening in life 'unbidden'; with energies with affects beyond that of the human. In terms of our engagement with the material and metaphorical world we inhabit, this perspective emphasises process and offers a way of reasoning individuals' participation in, and of, space. 'Spacing' is the unsettling and momentary resettling of things, feelings and thoughts, in the eruption of energies and in the vitality of energy and the interstices between things. It is in such liveliness that creativity, for them, emerges.

'Spacing' emphasises capacity and energies for change; abrupt or steady, non-linear and non-accumulative, discontinuous and held on to, uncertain; the unexpected and unbidden, influences that are other-than-human (Crouch, 2003a).

By engaging human activity in a wider world Deleuze and Guattari expand the complexity, multiplicity and possibility not only of things and [their] vitality, including humans, but also the ways in which we might address, consider and enfold things. Using this perspective, we can see how a lay geography is created, through a practical ontology of feeling, doing and thinking. It is emergent in spacing, a process that is kinaesthetically sensual, inter-subjective and also extra-human in affective constitution, expressive and poetic. Creativity is thus informed through combinations of different times and life durations and rhythms, different registers and intensities of experience. In particular the poetic process of spacing continually emerges, creative, contingent, awkward and not blocked in repre-sentations. It is through this process that the constitution of lay geographies is profoundly creative.

Creativity in community gardening

The creativity of gardening is partly in its material soft collision, reminiscent sometimes of contemporary art of 'found objects'.

> [He was] fascinated by the surroundings of allotments. Nothing wasted – all recycled – pots and pans, baths, old enamelled advertising signs, bedsteads, all had a use. Sheds made from old doors and window frames, haphazardly nailed together fused with vegetables and flowers, weeds, puddles and bare earth ...
>
> (Crouch, 2003c: 51)

Another part of the material expression of creativity is in the patterning of the ground, the way the ground is marked in particular cultivation distinct between people with more and less time to spend, with differing attitudes towards order or naturalness as much as ethnicity.

Combined, these different components constitute a creativity of an allot-ment aesthetic partly born of an attitude of recycling, partly of a lack of time and attitude towards order. Instead, freedom has accompanied much allotment cultivation, or allotment holding, for some decades. While allotment holding has a history of regulations and controls, it is in the patterning of the ground, the use of materials, feelings and human interaction through which creativity happens and grows over the edges of any surviving controls. The material space combines with a metaphorical space: attitudes towards nature, freedom and control, and in the way the ground is worked and cared for. In this combination, as much serendipitous as programmed, the vitality of spacing is evident in the constitution of space.

The artist Peter Lanyon identified a similarity between his rhythms in painting and those he felt in gardening. These rhythms of movement pattern the allotment aesthetic; they create its landscape of work and expression in movement. The spaces work relationally with the plotters' lives in the mutual and multiple flows of cultivating the ground. Gardening involves mobility, gentle and caring, strong and effortful: looking after small plants, lightly putting a hand up beneath netting to pick raspberries, dragging a mound of leafy growth to a compost heap. These practices each have a distinctive rhythm.

Even though Deleuze and Guattari downplay the body in their work, as too subjectively focused in a world of much more fragmented activity, it is difficult to ignore the embodied activity in this creative practice. One plot holder, Carol, demonstrates the close physical significance and making of values that she associates with allotment cultivation, simply in feeling her way around what she grows, how and where she grows it:

> [W]orking outdoors feels much better for your body somehow ... more vigorous than day to day housework, much more variety and stimulus, The air is always different and alerts the skin, unexpected scents are brought by breezes. Only when on your hands and knees do you notice insects and other small wonders. My allotment is of central importance in my life. I feel strongly that everyone should have access to land, to establish a close relationship with the earth ... essential as our surroundings become more artificial.
>
> (Crouch, 2003a: 1953)

Carol describes this activity through what she does, through how her body engages the intimacy of space, movements among multiple spaces between vegetation, earth, insects, the air and herself. She is closely aware of the effect of nature in what she does. Yet she is coping, getting by, and finds things happen along the way, as does Harriet, who lives on the outskirts of Manchester UK:

> I have lived in flats all my life and currently live on a council estate. I have no hope of ever being able to afford a garden, since my work is rather low status and underpaid. My allotment has enabled me to find a side of myself I did not know existed and it also helps me to cope with an extremely stressful job in a stressful city.
>
> (Crouch, 2003a: 1951)

In her coping she finds 'another side of herself'; her gardening has been a creative act, one of discovery and innovation of the self. In so doing she creates a temporary world, identity and feel of belonging: she discovers different things about herself; she acts creatively.

Yet these instances express something more than merely coping. The individuals are emerging differently through what they do, the way their life enfolds into and through the practice that becomes performative in opening beyond itself. Working their plot of ground, a particular space, they find that, often unbidden, something else happens: they become, or feel that they are someone different. In between action, in the gaps of things, individuals can feel becoming. Jimmy from South Shields put it like this:

> Any depressions that you might have, they just physically go within minutes of you getting down here [the plot] and sitting down. I think you're away from the home environment – not that we don't get on at home, but it's a fact

that it's nice to come away; if you want to come away, and have your own little bit kingdom. This is my little bit kingdom.

(Crouch, 2003a: 1955)

Creativity like this also occurs inter-subjectively. Linton is a plot holder in Birmingham UK. He talked about the social activity of giving crops away as part of the enjoyment of growing, cultivating and friendship:

If you give somebody anything they say – where did you get it from – I say I grew it myself. You feel proud in yourself that you grows it, you know, if you get it from the shop, some of it don't have any taste. On the allotment you plant something and it takes nine months to come to.

(Crouch, 2003c: 15)

He finds that through his gardening he comes into fresh contact with his world. What is given, or shared, is not merely to have something to give, it is the fulfilment, for a moment, of what he has been doing over time, the cumulative result of his performance of this plot space. What he has grown through the season becomes an object of social encounter: love and care of his ground, the way he works it, the person he engages in the gift, as well as developing non-commodity values and self-identification. A similar kind of becoming is marked by Deirdre, from Northampton: 'My allotment means relief from stress, creativity, the love of creation ... it means everything to me. I don't know where I would be without my garden' (Crouch, 2003a: 1956). Individuals suggest the transformative possibilities of the simple, uneventful things they do in terms of feeling rather than outcome. The creativity is not so much in the patterns on the ground but in what those patterns signify: changing relationships, feelings, ways of being and becoming.

In each case there is an acknowledgement of creativity being worked for and worked at, as well as arising out of what has been done. What they do and feel is enacted in relation to spaces. Space takes on or is given new significance in a process of spacing, like the man enjoying the respite of sitting in his shed, knowing he is surrounded by his plot where he has been working. Their creativity emerges from doing and exceeds merely getting by, producing some carrots.

Similarly, two plotters, Winifred and Adam talking, express values of ethnicity and cultural exchange nurtured in cultivation:

A: We learn from each other. You are very social and you are very kind. You make me feel good, you don't come and call at me. Things in your garden you always hand me, little fruits ... which I have valued so much.

W: And I've learnt from him. I've learnt some ways of planting, I've learnt real skills about planting, Jamaican ways of growing and cooking ... And I've also learnt about patience and goodness and religion, too. It all links in.

(Crouch, 2003a: 1952)

Through working their own plots in quiet close proximity over a number of years, they discovered more in the world and in themselves in a process of trying out, adjusting, improvisation and coming across new ways of doing things in hybrid combination that also reveal things new about themselves. Process occurs not merely in the sharing but also in the bodily actions associated in its follow-through. There is a mutual and multiple intimacy and complexity of nuanced mobility and habitual rhythms in this work, events and participation.

Carol, in the quote above, also draws attention to the way in which nature is creatively reconstituted in performance: her relationship with the earth, her doing body. This character of 'body in action' informs what each of the individuals makes of the performance, discursively and pre-discursively. She makes new space through her doing, not as an object 'out there' or merely 'felt' through the body, rather as constituted of numerous feelings and sensualities. Radley (1995) points to the significance of expressivity in the way individuals do things, and thereby the way in which performance is felt in a relationship with others, including objects and space, and things are changed. Expressivity is creative, emerges in the encounter, in the performance, the body-performativities in touch with other sensibilities, body-thinking and feeling. As Deleuze argues, 'all begins with sensibility' (in Harrison, 2000: 497). In the act of performing they are there, at the moment, doing, borne on a longer duration of action. Doing different tasks, each bears its own rhythmic character. Carol's performance is significantly unbidden and diverse; she seeks to cultivate and discovers other things; it happens. Her notions of nature are unsettled, enlarged and focused in the performance she makes. She builds, refigures and reworks nature through her lay knowledge, in a creative act.

In gardening, nature can become creatively refigured. It is rendered sensuous through the ways in which the individual may work the ground, encounter an idea of nature and come across stuff: beetles, worms, earth and stones, leaves wanted and unwelcome. New natures are creatively constructed. Through imagining and feeling 'nature' in its materiality, ideas of self and human relations with it, or in it, are refigured in the energies that flow amongst the numerous participants, rhizomatically but with attention to roots too. While there is evidence in what community gardeners say of their creative discoveries, things anew, there is also evidence of creatively (re)discovering a means to hold on to their world, their identity, to whom they feel they are and to what matters to them in their lives.

In 1998, a different creative performance was animated through a gentle ecological politics in a community garden at Uplands, Birmingham UK, in an event called Bloom 1998 (Crouch, 2003b). Performance and installation artists spent six months at Uplands interpreting what community gardening meant for people with plots, and combined this with their own ecological agenda. Over one thousand people attended the culminating performance over one evening.

The performance was situated very explicitly in a discourse of human, social and cultural/political relations. Uplands was chosen not only because it is a large site and spread over a hill, providing a dramatic setting for spectacle, but because the plot holders come from many different cultural backgrounds, displaying diverse ways of cultivation, ethnicities and a significant level of associative activity that involves the local community. To a degree, this space delivered an open stage and space that provided a context for performing a number of ideas through the use of 'natural' objects – earth, branches, seeds, movements, silence and laughter.

In Umbrella Gardening, according to Martin Burton, artist, one plot was transformed into 'a field of umbrellas planted in the earth. Our expectations are confounded by a paradox of light and sound of rain'. Some one hundred umbrellas were 'planted' in the ground and lit from beneath as the evening progressed. This lighting was accompanied by the projected sound of falling rain that had an effect of sounding also like roots moving, growing in the earth. The idea of the silent growth of plants was confounded. In another part of the event one hundred plastic gloves were 'planted' in the ground of one plot, raised up like their plot holders struggling with their land, and with their rights to 'hold' land – but also to unsettle the idea of growing nature. And of course this chimes with the history of allotment holding as one of struggle for rights to use land (Crouch, 2003b). A third part of the performance was presented by Blissbody, who 'tended their allotment/instal-lation throughout the early evening, with a ritualistic performance at dusk': the artists lit fireworks and set light to planted twigs. Situated on the brow of a hill, this performance made a significant spectacle in the dark. In each case the staged event leant away from a direct use of objects yet was performed in relation to, and alluded to, ideas of land. Each took place surrounded by many other plots where crops were growing.

Bloom 1998 offers another way of understanding the creativities of community gardening. There was a purposeful creativity. In their creative 'artistic' interven-tions each of the artists sought to represent, and/or express, to 'present' what they grasped of the everyday embodied feelings through-practice-and-performance of the community gardeners. In community gardening creativity occurs in the effort of moving life forward. There are different nuances in these distinctive creativities in performance, and they each have different registers of intensity, reflexivity and 'arrival'. For the community gardeners this performativity was in everyday nego-tiations with life, through which emerged, or erupted, nuances of creativity in the handling of things and feelings. For the artists, this practice and performance of every engagement and encounter presented different opportunities and produced different affects, albeit affects that were embroiled with non-human affects. The working of these flows of influence and feeling is considered in the conclusion to this chapter.

Conclusion

Creativity is not pushy; it does not necessarily insist. The more explorative, uncertain and tentative ways in which our surroundings become engaged in living suggest a character of flirting, exemplified in the way one often comes across very familiar sites seeing new juxtapositions of materials and materialities, as it were, 'unawares'. The unexpected opens out; we discover new ways of feeling, moving and thinking, however modest these may be, unsettling familiar and expected cultural resonances and the work of politics. Encounters like this may happen in increasingly diverse and complex ways across multiple spaces and in the ways in which we engage with and in them. Even in familiarity and habitual rhythmic engagement, the meanings, and our relationship with things, can change in register; slight adjustments of feeling over time becoming more significant.

The mid-twentieth-century artist of the International Modern Movement Peter Lanyon wrote of his sensitivity to rhythms in his acts of painting and in his gardening. In other writing he wrote not only of the acts of wielding a brush but of the experience of wandering around cliffs, hills and fields, of sitting on top of a bus and driving his car: moments where he found inspiration in the spaces he was moving through, at different paces (Crouch and Toogood, 1999). These resonances between artwork and 'simply' doing gardening, cultivating a piece of ground, suggest that, without flattening different types of skill or their end result, creativity can be understood more inclusively with regard to the mundane.

> I am able to take pleasure in my walks and in my gardening and to accept the factual sort of existence of stones and things because I see them as my source. I find with this a bodily rhythm in digging and moving (the sort of rhythm I used to feel through the seat of a car when driving very fast.) and I feel this produces a rhythm in my work.
>
> (Lanyon, letter to Naum Gabo, 1949, in Garlake, 1995: 233)

The commonly presumed distinction between 'art-creativity' and the emergent creativities performed in everyday life dissolves, at least in a way that exceeds particular learnt skill and competences. In each incidence of emergent creativity there is an active negotiation and a chance of 'moving on' with things, with life. Just as what is acknowledged to be artwork confronts, handles, and finds possibility, so does what community gardeners, and we ourselves, as individuals, do in the moments of lives.

In this chapter I have urged a reconsideration of the way in which cultural creativity tends to be thought, considered and contested. I have sought to open up the ways in which creative energies inhere in everyday living, in the hidden, and possibly unbidden, quiet excitement of the performance of mundane life. Hallam and Ingold undramatically articulate this kind of living, a slow emergence of the process of creativity in mundane performativity and its spaces. They reveal that everyday life can liberate creativity rather than repress it. Yet at the same time Deleuze and Guattari assist in further unpicking of the process, for they articulate how in the calk, slow and quiet, lies *intensity* in the sense of feeling, connectivity, and (despite their distrust, or avoidance of the notion) meaning.

Excitement in life may, after all, be quietly emergent, yet no less excitingly, in just such simple acts as community gardening. It is these subtle yet complex performativities that cultural geographies have found difficult to acknowledge. To point out this neglect is not romantic; to avoid it in the pursuit of the more easily intense is.

11 Growing places

Community gardening, ordinary
creativities and place-based
regeneration in a northern English city

Paul Milbourne

Introduction

> Cities are cauldrons of creativity. They have long been the vehicles for mobi-
> lizing, concentrating, and channelling human creative energy. They turn
> that creative energy into technical and artistic innovations, new forms of
> commerce and new industries and evolving paradigms of community and
> civilization.
>
> (Florida, 2005: 1)

Much has been said during the last few years about the importance of the cultural
industries to the renaissance of cities in various countries. Prompted by the writings of Richard Florida (2002; 2005), attention has been given to the economic
basis of creativity, the preferences of the creative class(es) for particular types of
place and the specific forms of urban regeneration associated with such creativity.
More particularly, the role of artists in the gentrification of disadvantaged parts of
the city has been discussed. The colonisation of low-rental ex-industrial spaces
in different cities by groups of artists has been viewed as an important part of the
urban regeneration process, producing new creative clusters, reinventing previously marginal spaces as 'buzz' places and attracting both property developers
and new population groups into the city (Hubbard, 2006; Zukin, 1989; 1995).

In this chapter I want to extend this discussion of the linkages between creativities
and the regeneration of disadvantaged spaces in the city by exploring the activities and
impacts of a community gardening group in a city in northern England. The spaces
within which this group operates have not, to date, proved that attractive to the creative classes or property developers. The place that I want to discuss is also residential
in nature, consisting mainly of terraced housing, and thus does not really offer the
types of working (warehouse) space that previous studies suggest is typically sought
by creative groups. In addition, there appears to be an absence of some of the social
and cultural attributes of place that it is claimed are sought by the creative class. For
example, the area is largely white and working class in social make-up, and continues
to be characterised by population decline and a range of social problems.

My focus on community gardening in disadvantaged urban spaces also provides
a different perspective on the relationship between creativity and regeneration in
the city. While gardening has long been discussed as a hybrid form of creativity,

I want to suggest that community gardening in places of poverty is producing different, but equally important, vernacular forms of creativity that are contributing to the reinvention and, in some cases, the regeneration of these places. In the chapter I will highlight how community gardening projects are able to transform the social and cultural, as well as physical, attributes of space and, in so doing, remake place and create new forms of sociality and conviviality. The empirical focus of the chapter is a small-scale community gardening project in Salford, a city in north-west England. A key aim of this project has been to green the back spaces that lie between the terraced houses in a disadvantaged area of the city. As such, the group is working within spaces that can be labelled as everyday, ordinary and mundane; the types of 'lived-in spaces' that some have claimed have remained marginal within urban geography (Hubbard, 2006). My work with this gardening group has involved in-depth interviews with members of its coordinating group, mobile discussions with other participants in and around their gardens, and three periods of participant observation at key events in the group's gardening year. Drawing on observations from this work, I will discuss the roots of the project, the spaces within which it operates and the cultures and creativities bound up with its activities. Before doing this though, I want to position this case study of community gardening in Salford within a broader context of published work on the cultures of ordinary spaces and environments of poverty.

Ordinary spaces and environments of poverty

It is has been claimed recently that the ordinary, everyday and mundane cultural spaces of the city have been largely neglected by geographers (Hubbard, 2006), with Miles *et al.* suggesting that the 'city of people, of everyday life, common occurrences, small shops, bus stops, allotments and waste ground is every bit of 'the city' as the more visible and high profile' (2003: 257). One of the first writers to point to the significance of these ordinary spaces of culture was Williams (1958b). In an early essay on the ordinariness of culture he argues that our familiar landscapes have been shaped by the everyday actions and associated creativities of ordinary people. Williams's interpretation of the cultural landscape has been drawn upon by cultural geographers (see Jackson, 1989), with the claim made that culture is ordinary precisely 'because it insinuates itself into our daily worlds as part of the spaces and spatial practices that define our lives' (Mitchell, 2000: 63). Moreover, if we accept that the culture of cultural geography is also 'a culture of everyday actions and social structures, a culture that humans mould through conscious and unconscious actions' (Groth, 1997: 11), then ordinary spaces need to be awarded increased significance within cultural (geographical) studies.

Writing thirty years ago, Meinig (1979) makes a powerful case for researching ordinary spaces (or landscapes, to use his terminology), seeing these as the geographical equivalent of the social histories that have revealed much about the previously hidden lives of ordinary people. For Meinig – and others who followed this approach – ordinary landscapes constitute the 'continuous surface which we can see around us':

[the] parts of an ensemble which is under continuous creation and alteration as much or more from the unconscious process of daily living as from calculated landscape design. Insofar as we focus on particular landscapes, we are dealing primarily with vernacular culture.

(1979: 6)

These ordinary spaces are also associated with a great deal of cultural meaning and modes of social behaviour, reflecting 'individual actions worked upon particular localities over a span of time' (ibid.), as well as complex spaces of vernacular creativity.

Within this chapter I want to focus on vernacular cultures and creativities in the context of environmental projects in disadvantaged urban neighbourhoods. During recent years increased attention has been given to environmental themes in disadvantaged places in the UK. Drawing on discourses of environmental justice, the hazardous nature of local environments in such places has been highlighted, particularly their proximity to sources of pollution (see Walker and Bickerstaff, 2000; Friends of the Earth, 2001). In addition, recent place-based studies of poverty have begun to highlight the ways in which the degraded state of the local environment contributes to processes of social exclusion in places of concentrated poverty: it reinforces the negative stereotyping of place, reduces quality of life and compounds residents' feelings of disempowerment (see Burrows and Rhodes, 1998; Lucas *et al.*, 2004; Lupton and Power, 2002). As Warpole argues, 'disadvantaged communities often get penalised twice. Not only do they have to live with fewer economic resources, they ... almost always live in environments which exact an additional toll on their well being' (2000: 9).

While these studies have developed useful understandings of the environmental components of disadvantage in poor places, they can be criticised for accentuating the negative attributes of these places. A more positive approach emerges from recent research on the impacts of community-based environmental projects – what, following Warpole (2000), could be described as ordinary forms of environmentalism – in disadvantaged places. In addition to improving the aesthetics and functionality of physical space, these projects have been shown to have produced broader social and cultural impacts through, for example, increasing levels of social capital and well-being, improving residents' sense of place and encouraging civic participation within local public space (see Burningham and Thrush, 2001; Church and Elster, 2002; Groundwork Trust, 2001; Kitchen *et al.*, 2006; Morris and Urry, 2006). What also emerges from recent work on this subject is the range of creativities linked to these types of environmental project, involving the formation of new forms of generosity, conviviality and sociality. In addition, artistic creativities are evident within these projects, as people fuse environmental and cultural identities in an effort to reflect personal and collective understandings of place. The next section of the chapter explores these sociocultural dimensions of local environmentalism in greater depth through the case study of community gardening in the city.

Growing places: community gardening in marginalised urban spaces

Gardening has long been recognised as an activity that extends well beyond the horticultural to embrace sets of cultural and creative themes. It is claimed that gardens provide fascinating everyday socio-natural spaces that are packed full of complex meanings and social actions (Bhatti and Church, 2001). For Longhurst, gardens represent 'paradoxical, ambivalent and equivocal spaces that reflect and embody a fusion of possibilities' (2006: 589) – complicating the boundaries between nature and culture, public and private, work and leisure, and individuality and sociality. Gardens can thus be seen to constitute ordinary spaces of 'social interactions, encounters, meanings and cultural exchanges':

> gardens are experienced as a private retreat; a haven from the public world; a setting for creativity; a social place for sharing; a connection to personal history; a reflection of one's identity; a status symbol; and as a natural world rendered more comprehensible.
>
> (Bhatti and Church, 2001: 380–1)

It is also been suggested that gardening represents a form of place making which involves the fusing of 'unmediated elements and processes of the physical world with human art and culture' (Crozier, 2003: 86) through everyday social interactions. For Crozier, not only does the garden need to be approached less as a 'place to get lost in' and more as somewhere 'to get found in', but the practice of gardening also involves a variety of symbolic performances that, collectively, lead to the creation of new senses of place (see also Seddon, 1997). As such, gardens constitute extremely creative spaces, reflecting not only horticultural skills but also different forms of art and design, including landscape design, the display of plant types and colours, and the selection of garden furniture.

Community gardening encompasses a large number of activities but is probably best defined as 'an organized, grassroots initiative whereby a section of land is used to produce food or flowers or both in an urban environment for the personal use or collective benefit of its members' (Glover *et al.*, 2005; see also Lawson, 2005). The community gardening movement has its origins in American cities of the late nineteenth century, where gardens began to be developed by low-income groups to grow food for local consumption on land that was assigned little market value (Schmelzkopf, 1995). In more recent times, community gardening has come to involve much more than collective horticultural activities aimed at producing cheap local foodstuffs. Most existing projects in the US have developed as a response to poverty, environmental degradation and the lack of safe green spaces in deprived urban places (Holland, 2004), with some also having their origins in civil rights struggles in the 1960s and 1970s (Ferris *et al.*, 2001). As such, community gardening comprises a broad range of horticultural, environmental, social and political concerns (Stocker and Barnett, 1998), combining 'the best of environmental ethics, social activism and personal expression' and involving 'a faith that what they [the gardeners] do not only helps the individual but strengthens the community' (Lawson, 2005: 301).

Recent studies in the United States identify the power of community gardening to transform place. In addition to providing their members with a supply of cheap, fresh and locally produced foodstuffs, communal gardening has also produced new forms of sociality, conviviality and meanings of place. These are very much place-shaping and place-making projects: they involve the physical transformation of despoiled spaces into more functional and aesthetically pleasing spaces; they develop new communitarian and participative spaces, within which members of the community come together to participate in gardening activities; they also counter the negative stereotyping of disadvantaged places and create new senses of identities among residents.

It has been suggested that community gardening projects create new forms of 'third space' (Bhabha, 1994), which is positioned 'outside of work and home where people can gather, network, and identify together as members of a community' (Glover *et al.*, 2005: 79). In addition, these third spaces of community gardening lie between the public and private realms. As Schmelzkopf (1995: 379) comments, they 'are part of the public domain and are the sites of many functions conventionally equated with the private sphere. Domestic activities, nurturing, and a sense of home are explicitly brought outside into the gardens'. Recent studies also highlight the sociality, reciprocity and trust associated with community gardening, involving the fostering of new forms of sharing and generosity, informal economies and civic culture, with gardening often acting as a springboard for the development of broader, place-based social and political actions.

The community gardening agenda in the UK has become more significant during the last couple of decades, due in part to the actions of two national horticultural organisations – the Federation of City Farms and Community Gardens (FCFCG) and the Royal Horticultural Society (RHS). The FCFCG is a voluntary organisation that supports, represents and promotes community farms and gardens in the UK, with many of these based in disadvantaged urban places. The RHS has been actively promoting local community gardening projects in the UK for the last couple of decades through its 'Britain in Bloom' campaign, which aims to encourage local communities to develop communal gardening projects that meet a range of horticultural, sustainability and social criteria. More recently, it has developed an 'urban regeneration' category and a 'neighbourhood awards' scheme in an effort to increase participation from communities in disadvantaged urban places.

Apart from noting the increasing number of projects, it is difficult to say much more about community gardening in the UK as little research has been undertaken on this topic. A survey of FCFCG members by Holland revealed that most community gardening projects are bound up with a broad range of environmental, social and community interests:

> the results showed that there was a multiplicity of purposes for their existence which related more to their function in community development, the schemes thereby acting as 'agents of change' … It would appear that what is grown

is secondary to what else is achieved, in many cases, and even where food growing is the stated aim ... other objectives are also achieved.

(2004: 303)

But this study did not get 'inside' these projects to explore the types of social and cultural themes identified by the US studies. Other research in the UK has been conducted on a different form of communal gardening – that undertaken on allotments – and here we do see evidence of the cultural dimensions of collective gardening. Drawing on findings from a national study of allotment gardening in the 1980s (Crouch and Ward, 1988), Crouch (1989) suggests that 'the culture in which the allotment grew up and was sustained was one of working-class agitation for improved conditions, and self-help' (ibid.: 191). Representing pockets of land, usually on the edges of residential areas, and regulated by local authorities, allotments are shown to be characterised by a cooperative culture of gardening, which involves cooperative forms of organisation and ownership, and the strong reciprocal relations between gardeners (see Crouch and Ward, 1988). In Crouch's words, allotment gardening provides 'fulfilment within a particular shared experience, a particular culture' (1989: 191).

Community gardening in a northern English city: place making and creativity in the back spaces of Salford

For the last couple of years I have been working with a community gardening group in the Seedley and Langworthy area of Salford in Greater Manchester. The selection of this gardening project as a case study for research on community gardening in UK cities was made for 'objective' and personal reasons. In 2006, this group was awarded first prize in the 'urban regeneration' section of the RHS's Britain in Bloom awards and, in one of my first interviews with the RHS, it was suggested to me that the Salford project would make an interesting case study for my research. Seedley and Langworthy is also a part of Salford that I know well. It is the place where my grandparents and other relatives lived and so was regularly visited during my childhood years in North Manchester. My intimate knowledge of this area also provided me with an insight into the area's downward trajectory during recent decades.

Seedley and Langworthy developed as a residential area of terraced housing in the nineteenth century to accommodate the growing number of workers employed by Salford's manufacturing industry. It is situated about two miles from Salford docks, which linked local manufacturing companies with the port of Liverpool via the Manchester Ship Canal. While most of the manufacturing industry has long disappeared, the nineteenth-century back-to-back, two-bedroom terraced housing remains. Described by Greenwood as 'jungles of tiny houses cramped and huddled together, two rooms above, two below' (1933: 13), these terraced properties also lack gardens at their fronts and rears. Since the mid 1990s the area has suffered from a range of problems, particularly population decline, the collapse of its housing market, which has included the abandonment of properties, and problems with anti-social behaviour. This downward spiral of disadvantage has also involved the deterioration of the local environment, the closure of retail outlets in the area, and declining levels of social capital, civic participation and quality of life. A local community organisation describes the area in the following terms:

Unemployment rates in the area are higher than in Salford and national averages, with particularly high rates of youth and long term unemployment. The area also suffers from high crime rates, with burglary and juvenile nuisance identified as specific problems. With the addition of low incomes, poverty and debt through negative equity, these issues compounded to create an area of isolation, fear and deep-rooted social exclusion for many individuals.

(Seedley and Langworthy Trust, 2009)

In 1998, Seedley and Langworthy was awarded almost £14 million of central government regeneration funding for the period 1999–2006. Much of this funding was targeted at housing market renewal, particularly bringing back into use those properties abandoned by residents. Other funding was aimed at schemes for children and young people, community safety initiatives, employability projects, health and well-being initiatives, and environmental schemes. In addition, Urban Splash – an urban-based property development company – has recently become active in the area, converting many rows of nineteenth-century terraced housing that were originally marked for demolition into contemporary living spaces, with these being marketed at young professionals.

The Seedley and Langworthy community gardening project emerged out of two of the streams of regeneration funding – community safety and environment. In 2002, an alley gating scheme was initiated as part of a crime reduction programme. The alleys – narrow spaces that run behind the rows of terraced houses – were widely recognised by residents and the local police as problematic spaces: they were dumping grounds for rubbish, places of anti-social behaviour and spaces through which properties were being broken into (Figure 11.1). The alley gating scheme involved the cleaning and gating of these alleys by the local authority, with access restricted to the residents of adjoining properties.

The community gardening project commenced with the first alley to be gated. One resident in this street, Jenny, who was a keen potted gardener, decided that she wanted to green her newly gated alley:

We had this sort of closure party, because it was the first one ... and because we've always planted all our life, me before I got married, same with my husband, I said let's put hanging baskets on the side there, look rather sweet ... I'll put a few pots down there. One or two [neighbours] were interested. So we just put a few pots like that with a few pansies, because it was late on in the year, couldn't do anything else. Just put them down just to fill it up a little bit. Moved a few pots out of our front for this part of it we had, it was a nice day, big buffet outside, and they said, 'Oh, could do this, could do that', it set everybody off.

(Jenny)

Figure 11.1 Back spaces

Tony, one of the community wardens funded through the regeneration programme, heard about the greening of this alley and made contact with Jenny and her neighbours. In his previous post in Manchester, Tony had encouraged a group of residents to develop a community gardening initiative through the RHS's Britain in Bloom initiative. He organised a meeting with local residents of the newly gated alleys and secured their agreement to develop a community gardening project in these spaces. In 2005, the residents took control of the project and began to develop it as a genuinely community initiative:

> So we've done two years and then this is our third year, as a community, as us in the community, because that's where I truly believe it should be. If you want to get the community involved and the people involved, then it belongs to the community, it doesn't belong to anybody.

> (Jenny)

The governance of the project reflects this community ethos. It is coordinated by a small group of local residents, all of whom are women, who meet on a monthly basis in the community one-stop shop. These meetings discuss the funding of the project; they also involve decisions being taken on the creative dimensions of the group's gardening activities, as is explained by Doreen: '[the meetings involve] choosing what plants, what colours to put in the troughs, what colours to put on the lampposts. Then there's the majors on the park wall, the Cornerstone [local community centre], we pick out what colours.' The creativities of the project stem from the motivations and commitments of this group. What emerges from interviews with its members is that this project is constructed as a way of re-creating the 'Salford of old', of reclaiming the front and back spaces of their properties, of rekindling a past community spirit, of making a difference. Some of its members are suspicious about external involvement in *their* project, given the history of both non-intervention and perceived inappropriate interventions (such as the proposed demolition of whole streets of houses) by public bodies, including the local authority, in the area. This is a project, then, that is being led by local people on behalf of local people, and the creativities that it has produced reflect this community philosophy.

Decisions about individual alleys are taken by neighbourhood groups, each of which is responsible for a small number of streets. Some initial infrastructural funding was provided for the newly gated alleys to repair hard surfaces and provide outdoor furniture. The gardening group also managed to secure limited funds to purchase hanging basket kits and, working with the local Youth Offending Team, to fix these onto the fronts of residents' houses on participating streets. Following this early work, the group has provided advice to residents on 'greening' their streets and alleyways, and organised the annual entry into Britain in Bloom. In recent years it has spent about £20,000 per annum on its activities, money that has been raised through applications to different funding bodies and the organisation of a variety of fund-raising events. Residents in participating streets also make a small financial contribution of £2–3 per month towards the horticultural upkeep of their streets and back alleys.

The Seedley and Langworthy gardening project has made a significant contribution to the regeneration of the area. While its origins lie in a government-funded programme of regeneration – which provided the resources for the gating scheme, some of the initial alley gardening infrastructure and the community warden who encouraged residents to develop the project – this community gardening project demonstrates the power of ordinary forms of environmentalism. The back alleys and the streets of this area have been physically transformed by the actions of this determined and ambitious group of community gardeners. Hanging baskets and containers of flowering plants can be seen on all streets and at all times of the year, and some residents have sought to develop impressive displays of evergreen and flowering plants outside their houses. In one case, recycled chimney pots have been used as plant pots in an effort to reflect the industrial heritage of the area. The project has not been without its problems, but the determination of certain individuals to carry forward the gardening project has enabled it to make significant progress in changing local perceptions of everyday spaces:

Figure 11.2 One of the alley gardens

We were always frightened of putting plants at the front because they would be pinched, which, at first, when we first started they were ... but I wouldn't be beaten and if that happened I went the very next day and bought [another] one and replaced it. [And how is it now?] They don't do it any more. I suppose it is because our street looks nice and some people are jealous of it.

(Vera)

Walking through the gated, furnished and planted back alleys is a moving experience, at least for me. These previously residual spaces have been transformed into spaces of nature, sociality, culture and creativity (Figure 11.2). They have become very much 'third spaces' that complicate the boundaries between the public and private arena, as well as individual and collective actions. The alleys now provide communal spaces for neighbours to socialise with each other, with acknowledgements in the street being replaced by meaningful conversations and the development of long-term friendships. These comings together are thus producing new forms of sociality linked to reciprocity and trust, as well as conviviality and a sense of shared belonging. The alleys have become spaces for the development of hope and of alternative futures within and beyond their walls, involving the cultivation of fruit and vegetables in some of the alleys, the organisation of food-based social events, the increased involvement of children in communal activities and, most significantly, increased involvement in the broader regeneration of the area.

These back alleys can also be viewed as important spaces for identity formation and creativity. The casual observer looking in through the gates gains only a partial impression of the alleys. As Karen states in relation to her alley, 'people only see the end of it but they don't see the middle of it and once you walk into

Figure 11.3 Personal spaces

it, it is like a secret garden'. While the alley gardens all appear very similar to the casual observer, on the inside of the gates the individuality of each alley becomes apparent – a reflection of the physical attributes of the alley (such as size and aspect), the collective gardening actions of each gardening team and the individual efforts of residents to personalise the patch of ground adjoining their homes. Flowering plants displayed in hanging baskets and ground containers are a feature of all the gardens, with fruit and vegetables also present in some, but the styles of planting, varieties of plants and types of container used vary enormously from one alley to the next. Creativity and meaning are particularly associated with the spaces adjoining some people's back doors. Here, not only are planting schemes more individual, but personal items have been introduced to give the space more meaning. Such items include distinctive pieces of garden furniture, specially designed pots and troughs, and decorative plates from different countries of the world visited on holiday (Figure 11.3).

The cultures of this community gardening project are also evident in other ways. The project works with local primary schools in an effort to involve younger members of the community in greening their local area. It has brought in local artists and poets to develop cultural perspectives on gardening through, for example, the construction of murals that are displayed during RHS visits. In 2008, the RHS encouraged gardening projects across the UK involved in Britain in Bloom to make connections with their local histories, through the theme of 'roots'. The Seedley and Langworthy project worked with a resident artist at the

Figure 11.4 Cultures of nature

Lowry Centre, a large arts complex in the recently regenerated docklands area of Salford, to develop this theme with local schoolchildren. 'In Bloom' was turned into 'In Bloomers', with the children producing their own representation of gardening within the shape of a pair of bloomers printed on a white tea towel. Many of these were then hung on clothes lines across the alleys as they would have been on washing days in the past (Figure 11.4). In addition, the coordinating group printed a series of images on bed sheets, which were also hung across the alleys, to highlight the roots of the area. These included prints of old street names that reflected the area's links to the docks, historical pictures of street scenes, and photographs of previous prominent local residents, one of whom was Emmeline Pankhurst, founder of the suffragettes – an early twentieth-century movement dedicated to securing votes for women.

In 2003, 2004 and 2005, Seedley and Langworthy came first in the North-West Britain in Bloom 'Urban Regeneration' awards. In 2005, it was positioned second in the UK 'Urban Regeneration' awards and in 2006 the project came first, as well as receiving a Gold Medal from the RHS. The judges commented that:

> One of the main benefits of Britain in Bloom is the regeneration of disadvan-
> taged areas and nowhere is this better seen than in Seedley and Langworthy

in Salford. Here the people have been given the opportunity by the Local Authority through selective funding of specific projects to improve their environment, making it safer, cleaner and greener. No vast sums of money could achieve the results seen here. Once derelict streets have been transformed by gating them off, adding floral and sustainable planting and letting the people develop a village type community spirit within the city. As judges, it gave us great pleasure in seeing at first hand what can be achieved by being a finalist. There is no better example of regeneration in action, a well deserved Gold Medal Winner.

(Personal communication, 2006)

What are evident in these comments are the linkages between the project, local regeneration and place (re)making. While my work with the project remains at an early a stage, evidence emerging from my interviews with project participants and the participant observation indicates that community gardening is impacting on the ways participants and the wider public socially construct and interact with their place. In 2006, the local community trust undertook an evaluation of the project through a survey of local residents. Key findings from the 200 questionnaires returned point to a range of social and community benefits associated with the project, with 86 per cent of respondents agreeing that 'the project has made me feel more positive about the area', 77 per cent concurring that 'the project has improved the look of my street', the same percentage considering that 'the project has helped bring people together', 63 per cent feeling that 'the project has made the area feel safer', and 80 per cent agreeing that 'the project has helped children and young people to get involved in their area'. The trust also conducted a participatory appraisal at a local community event, which indicated that residents most associated the project with collective action, community spirit and positive outcomes.

Conclusion

The spirit of creativity and social justice of famous past residents, such as Britain's foremost living classical composer, Sir Peter Maxwell Davies, and suffragette Emmeline Pankhurst, is reflected in Seedley and Langworthy's dynamic community today.

(Salford City Council, 2009)

My aim within this chapter has been to move beyond recent discussions of cities and creativity that have pointed to the centrality of the cultural economy and the choices and actions of the creative class(es), to explore more 'ordinary' forms of culture and creativity bound up with a community gardening project in a disadvantaged area of a northern English city. In doing this, I have highlighted how the actions of a small-scale and community-based environmental project, which has sought to introduce 'nature' into some of the area's residential streets and back alleys, have created new types of sociality, cultural identity and creativity within and beyond these spaces. In writing this chapter I have focused on the social and cultural aspects of this gardening project, in an effort to connect with the themes of this book. I could also have discussed the ways in which the project has created new forms of horticultural knowledges and political empowerment among its participants. Indeed, community gardening projects such as this one need to be viewed as a complex amalgam of the natural and the social, combining elements of horticulture, environment, community, culture and politics.

It is clear that in Seedley and Langworthy the creativities of the community gardening group have played an important part in the remaking of this place, altering the 'nature, social relations, and meanings' that Sack (1997: 84) considers to represent the constituent components of place. In growing plants, the group has also grown community, developing new meanings for gardening as well as for place. As Terry Whitehall, one of the judges of the project in 2006, stated, 'I was overwhelmed with Seedley and Langworthy's entry. In all my time judging this was the best display of a community coming together and community spirit that I have ever seen.' The group's activities also make connections with the area's industrial past and working-class cultures, through its local history, poetry and art outreach projects, blurring the boundaries between nature and culture.

The back alleys have been transformed from derelict spaces to socio-natural places, with the introduction of nature creating new social and cultural land-scapes. Within these previously back spaces, nature, sociality and culture collide in complex ways to provide extremely interesting 'third places' that are at least as significant for cultural geographers as the bookstores, cafes and coffee shops that Florida (2002) identifies as important spaces within the context of creative urban clusters in the United States. They too represent spaces that lie between the worlds of home and work, and the private and public arenas. They also represent highly symbolic spaces, filled as they are with a variety of cultural meanings, identities and creativities as well as, of course, potted plants.

> I went there for the judging, and I couldn't believe it. I have never experi-enced such a community in [that] every single person had the same story and they are all out talking to each other and having cups of tea on their tables that they had put in their alleys, saying I used to come home and be scared and run to my house and go into it and stay there, Now we spend our weekends out here talking to our neighbours. It's just incredible.
>
> (RHS Britain in Bloom co-ordinator)

12 Creative destruction and critical creativity

Recent episodes in the social life of gnomes

Tracey J. Potts

Following Jean Burgess's work on digital storytelling, I wish, in this chapter, to affirm 'as *part* of the contemporary vernacular the experience of commercial popular culture' (2006: 201). Focusing on recent episodes in the social life (Appadurai, 1986) of the garden gnome, precisely as a rich example of contemporary vernacular creativity, an extraordinary degree of creative action can be discerned which contradicts the received idea of gnomes as ready-made mass objects packaged with ready-made mass sensibilities (Dorfles, 1968). Gnomes, far from relaxing in their customary locations – beside ponds or on rockeries – barely seem to sit still; the existence of today's gnomes – 'liberated' by the Front pour la Libération des Nains de Jardin (FLNJ) or the Italian equivalent, Movimento Autonomo per la Liberazione delle Anime da Giardino (MALAG), sent on holiday by pranksters, invited into the metropolitan home and the design museum – can be considered to be quintessentially 'wild' (Attfield, 2000).

That said, part of the ambition here is to consider what is at stake in the creative act in the information age. With reference to Alan Liu's (2004) gloomy assessment of the possibilities for creativity given its requisitioning by a voracious digital corporate culture, I distinguish between creativities that are complicit with this ethos and those that offer resistance. A cold creativity founded in the production and accumulation of capital and 'good taste' can be discerned in relation to particular types of gnome fun, acting in contradistinction to a warmer, expansively *heterotopic* (Foucault, 1986) set of actions, relations and energies. To make this case I contrast the designer gnome (exemplified by Philippe Starck's gnome stools Attila and Napoleon) with a group of gnomes who live, against all odds, in a recycling plant in Birmingham, UK. Attending to the cautious choreographies that accompany the gnome in the context of contemporary design reveals not only that taste hierarchies are very much alive and well – contradicting the growing conviction that kitsch and so-called bad taste are 'dead' (Street Porter, 2001; Londos, 2006; Attfield, 2006) – but also that a symbolic economy is at work at the heart of the apparent relaxation of taste. It should become clear that not all gnomes are cool and much squeamishness can still be detected in relation to the gnome's presence. Meanwhile, in Birmingham those gnomes who find themselves remaindered – effectively sentenced to death by taste decisions – are brought back to life in the form of a gnome sanctuary.

Visitors to the site are greeted by crowds of gnomes and assorted garden ornaments, arranged in loosely bounded communities: a 'froggery' sits beneath a tree
clustered with owls, squirrels and other climbers. Drawing on Bruno Latour's
(1999) ideas around the constitution of sociality through things, the creativities
of staff at Lifford Lane are set against the socially anxious performances of the
'tasteful'. More, from Liu's point of view, these gnomes are constituted through
creative actions that resist capital interests and investments of 'cool': the site
at Lifford Lane can be seen, consequently, to offer an example of a 'strong art'
(Liu, 2004: 9), namely an art that is resolutely material, communitarian and at
odds with a world increasingly governed by perpetual innovation and relentless
refashioning.

'Ha-ha-ha, hee-hee-hee, I'm a laughing gnome and you can't catch me' (David Bowie, 1967)

For many years, garden gnomes have been held to represent the very opposite
of creativity; the received idea of gnomes is one of self-evident bad taste and
reified existence. Banned from the Royal Horticultural Society's annual flower
show at Chelsea (RHS, 2007: 11), viewed as 'archetypal kitsch' by arbiters of
taste (Dorfles, 1966: 4), gnomes stake out a particular geography of tastelessness
and exemplify a thoughtless and unreflexive expression of identity. In tracing the
relations between twentieth-century British and suburban identity, for instance,
David Gilbert and Rebecca Preston christen suburbia 'the gnome zone' (2003:
197): a piece 'guaranteed', in the mind of reviewer Alastair Owens, 'to produce a
fit of the giggles' (2006: 268).

More recently, though, the gnome has been subjected to an extreme makeover.
A recent Sunday supplement item summarises:

> Against all the odds, gnomes are being rehabilitated. Arguably, this began
> when designer Philippe Starck celebrated their kitsch factor with his Napoleon
> Gnome stool. But it didn't stop there. They feature in the Big Brother house.
> 'Garden gnomes are exploding all over America right now,' claims Elton
> John, currently writing the music for Nomeo and Juliet, an interpretation of
> the bard's classic, for Disney. And who has the BBC chosen to interest chil-
> dren in gardening? None other than Gordon the Gnome, with his catchphrase
> of 'let's get mucky'.
>
> (Siegle, 2005)

Gnomes, according to gnome biographer Vivian Russell, 'have quietly acquired
a cult following' (2005: 8). Starck is no exception in inviting the gnome over
to play. Sarah Lucas's *Gnorman* – a gnome swathed in cigarettes – acts up in
an American art gallery, 'thumb[ing] its nose at artistic values and refined taste'
(Bruce Museum, 2007: 2). Gnomes cavort with pop group Erasure in their promotional video for 'Make Me Smile', are used as a murder weapon in an episode of
television series *C.S.I.*, and over on the ABC Network, configure themselves into
an obscene message in *Lost* (telling Hurley's boss what to do with his job). The

gnome has, it seems, decisively left suburbia to take up residence with Prince Charles at Highgrove, visit Paris Hilton in Los Angeles (Fox News) and live underwater in the Lake District (Meredith, 2005). Gnomes help to merchandise shoes in Selfridges – the same gnomes are available for purchase as limited editions – offer themselves as 'fun' interior accessories at the respected British furniture retailer Heal's, and sell holidays for Travelocity.

Scenes from Jean-Pierre Jeunet's 2001 film *Amélie* distil the contemporary gnome situation: a cemented gnome marks the resting place of Amélie Poulain's mother's ashes and emblematises her father Raphael's blocked emotions and agoraphobia. Amélie's decision to send the gnome 'on holiday' – courtesy of a travel agent friend she arranges for the gnome to be photographed against various exotic backdrops – catalyses Raphael's recovery and re-engagement with life beyond his garden gate. After receiving a number of postcards from his gnome, the widower packs his case and sets off on his own travels. *Amélie* has prompted innumerable copycat incidents and gnome postcards have become a popular leisure activity, to the extent that a dedicated travelling gnome – A Gnome Away from Home, complete with passport – has been produced.

From the relative peace of the suburban garden, the gnome's contemporary situation is frantic, not to say fraught with danger. Gnomes are often sent away without consent and subjected to varying degrees of violence: kidnapped, hung, forced to commit suicide, stabbed and smashed (Reuters, 2000; BBC News, 2007). The very least that can be said is that the gnome, these days, is subject to intense *creative* activity and that his suburban past is behind him. Eva Londos, for instance, deduces:

> Another milestone that is passed on the way of rehabilitating kitsch is the present exhibition of people's garden gnomes at the prestigious House of Culture in Stockholm (Spa for Garden Gnomes). The garden gnome seems to have become respectable after centuries of exile from the sphere of good taste.
>
> (2006: 297)

The admittance of the gnome into high art/culture space is read as evidence of a cultural relaxation around questions of taste and value. Where once 'the upwardly mobile' cared about marking social distinction, 'anxiously ogling and copying the manners of the upper classes' (ibid.), today a more casual attitude is seen to prevail, to the extent that 'the concept of kitsch seems to be of little use and rather out of date' (ibid.: 299).

While it is certainly the case that gnomes are appearing in unpredictable places and enjoying a dizzying social life, to conclude that this gives indication of a generalised informality around aesthetic judgement is wishful thinking. Paying attention to the specific circumstances in which gnomes live or die – or in Victor Buchli's material culture view, are 'released, given away, wasted, taken away, sacrificed or disposed of' (2002: 16) – exposes a considerable degree of commotion, directly contradicting the vision of gnomes' freedom to roam 'without passports' (Londos, 2006: 299). The gnome may well be a 'wild thing' in Attfield's

(2000) sense of the phrase, but his or her role in the taste game is *convoluted* rather than obsolete or uninhibited.

Before paying a visit to particular groupings of gnomes and the actions by which they become constituted, it would be useful to pause, first, for theoretical elaboration (specifically the *work* of the aesthetic in the post-industrial era of information) and second, for a brief visit to gnome past. Once the mechanisms of creative destruction, symbolic economic production and investment are made clear, the political distinction between cool gnomes and resuscitated gnomes should become discernible.

Creative destruction and the technicians of cool

The question of creativity in the twenty-first century, for Liu (2004), is of utmost concern. Augmenting Joseph Schumpeter's ideas around the economic motor of capitalist modernity, Liu draws attention to the commandeering of creativity and innovation by business in the information age. If, at the beginning of the twentieth century, industrial capital drew energy from, to use Schumpeter's words, 'a perennial gale of creative destruction … incessantly destroying the old (structure/product/technology/organization), incessantly creating a new one' (cited in Liu, 2004: 2), by the time of the new millennium the storm has intensified and shifted ground. The momentum of the destruction/creation circuit in the post-industrial scene has not only accelerated, an aggressive takeover bid of the aesthetic has been tabled and, in the main, accepted. Liu expresses his concerns over the future of literature (and by implication its sibling arts) in the 'age of advanced creative destruction' (2004):

> The vital task for both literature and literary study … is to enquire into the aesthetic value – let us call it simply *the literary* – once managed by creative literature but now busily seeking new management amid the ceaseless creation and re-creation of the forms, styles, media, and institutions of postindustrial knowledge work … After all, next to the great institutional documents of our times heralding 'innovation' in their very logos – the legion of 'dot com' company prospectuses, websites, advertisements, and so on – what could literature be but a minor act of creativity, like a screensaver?
>
> (ibid.: 2–3)

With the information economy poised to dominate creative output (as the result of the 'colossal convergences' (ibid.: 3) inaugurated by the rise of networked ICTs and technoscience), the question of creativity – *the future literary* – outside of capitalist relations, remains vexed and urgent.

Creativities that run counter to the flow of information attempt to contradict the 'gigantic now' of modernity with its 'apparently seamless necessity of modernisation' and so-called progress. The greatest source of friction to the smooth operations of progressive, rationalised development – as Walter Benjamin's (2003) angel knows – is history. Liu puts it succinctly:

As arbitrageurs of 'creative destruction' might say: he's history. History – including the history of modernization itself (now identified with smokestack industries and ossified organizations) – is obsolescence.

(2004: 6)

Figured as impedance ('innovation means, first, the systematic sloughing off of yesterday'), history all the same retains the power of the other, 'the other of the future' (ibid.: 18): to disrupt and apply pressure to the dominant scene of rapid and ruthless innovation. History is, then, refigured by Liu as weight, as a material force: 'it is difficult today to think of any authority other than history that has the heft to match the ideology of postindustrialism and globalism' (ibid.: 376). History is the spanner in the works or the virus in the operating system. Not-for-profit creativity must attend to history and is hence a form of 'matter work' (ibid.: 43), labour consigned to the developing world in the era of informationalism:

> Where once the job of literature and the arts was creativity, now in the age of total innovation, I think it must be history. That is to say, it must be a special, dark kind of history. The creative arts as cultural criticism (and vice versa) must be the history not of things created – the great, auratic artifacts treasured by a conservative or curatorial history – *but of things destroyed in the name of creation.*
>
> (ibid.: 8, my emphasis)

Critical creativity must fix upon the scene of destruction and, with it, the stubborn materiality of development; the future literary necessitates the 'rejection of the aesthetic ideology of critical innovation ("make it new") in favour of an ideology of critical destruction'. Critical creative practices ('strong art') thus include 'acts of delay, displacement, oblique representation, and stylisation', acts designed to grate against, if not destroy, 'this eternal now' (ibid.: 9).

The elision of history in the consumption process helps found the dominant aesthetic of the information age: cool. Liu, again, elaborates:

> Cool is the techno-informatic vanishing point of contemporary aesthetics, psychology, morality, politics, spirituality, and everything. No more beauty, sublimity, tragedy, grace, or evil: only cool or not cool.
>
> (ibid.: 3)

Cool necessarily represses history and is in every sense governed by what Pierre Bourdieu would term a 'charismatic ideology' (1976: 11). To attempt to anatomise cool is absolutely uncool, as Xander Harris discovers in an episode of *Buffy the Vampire Slayer*:

Xander: But ... It's just that it's buggin' me, this 'cool' thing. I mean, what is it? How do you get it? Who doesn't have it? And who decides who doesn't have it? What is the essence of cool?

Oz: Not sure.

Xander: I mean, you yourself, Oz, are considered more or less cool. Why is that?
Oz: Am I?
Xander: Is it about the talking? You know, the way you tend to express yourself in short, noncommittal phrases?
Oz: Could be.

(1999)

Cool things are entirely without essence and, hence, call for technical management. In Jean Baudrillard's thinking, the modern order requires 'intelligent technician[s] of communications' (1996: 26). 'The will to design' becomes a species of knowledge work entailing the careful manipulation of information to ensure 'the perfect circulation of messages' (ibid.: 29).

Rendered as information, cool things constitute, irresistibly, the grounds of class formation. As Bourdieu reminds us, questions of aesthetics operate inescapably as 'a dimension of the ethics (or better, the ethos) of class' (1976: 47) and the ability to generate legitimate messages or aesthetic codes is inequitably distributed. Following Bourdieu, Beverley Skeggs notes how the aesthetic and its charismatic appreciation congealed in notions of taste is 'always defined by those who have the symbolic power to make their judgement and definitions legitimate' (2004: 107). With this, Liu (2004) notes the shapeshifting proclivities of white-collar professionals together with the rapid churning of capital that is the constituent feature of post-industrial class formation. Technicians of cool thus stand to make large profits and successful creative destroyers constitute a new social class. The material assets that comprise the personal portfolio of the possessive individual become less important than their shrewd manipulation and exchange (Skeggs, 2004). In a universe that churns capital – economic and symbolic – at bandwidth speed, the value of things becomes subject to sophisticated forms of asset stripping and revaluation. If, mid-century, respectability as a codeword for a middle-class bodily hexis (Skeggs, 1997) manifested in restraint, today the mobilisation of tastes, styles and objects in the name of class can appear bewildering in the extreme. Perverse, twisted tastes – the appreciation of the 'so bad it's good' and still more convoluted positionings – are now, in contradistinction to those who see the democratisation of bad taste, the markers of metropolitan sophistication (Potts, 2007). David Harvey's idea of 'accumulation by dispossession' (2008: 39) is instructive here in that the property of the 'other' becomes an attractive source of cultural value for the cool technician (see also Liu, 2004; Skeggs, 2004). The ability to expropriate culturally reviled objects and dubious tastes and to remobilise them as markers of sophistication – effectively to generate and extract surplus value from things regardless of their career biography – is, therefore, the new class imperative. More, in these post-ironic times, the more dubious the object's career, the greater its possibilities for refurbishment as an exemplar of 'edgy', cool style. In other words, it is precisely the gnome's prior existence as a code word for suburbia that *informs* its contemporary usage as design material. A brief survey of the gnome's biography should help to make this clear.

Gnome past

The authorised history of suburbia details the appearance of gnomes in the garden as a customary story of 'trickle down'. Following Thorstein Veblen's (1899) notion of the descending flow of style through the social order, commentators have construed the suburban garden in terms of its mimicry of Victorian aesthetic values. Helena Barrett and John Phillips view this positively, valorising the gnome's noble origins:

> Most of the deprecations of suburban gardens have been flung at the ubiquitous garden gnome placed in his rockery. Yet both gnomes and rockeries as garden features have perfectly respectable, even aristocratic antecedents, being emulations of examples found in stately homes that were, even in Victorian times, avidly visited by the general public.
>
> (1993: 177)

Others are less enthusiastic, seeing the logic of emulation as vulgar by reason of the sin of pretentiousness (Dorfles, 1968; Binkley, 2000). The taste for garden furnishings and accessories, which emulates aristocratic schemes, is, therefore, derisible as failure to distinguish between fantasy and reality. Gnome lovers, in short, have ideas *above their station*.

Worse still, gnomes begin to take their shapes from popular culture, resulting in what some see as an explosion of dubious taste. A Mass Observation file report 'Public Taste and Public Design' berates those who, having enjoyed Disney productions such as *Fantasia* and *Snow White*, succumb to a taste breakdown in the garden. Amid garbled ideas about the intellect of 'the masses', the author is clear about the significance of the gnome:

> All of Disney's films, indeed, show a sensitive and artistic touch which – by the rules and precepts of a dozen years ago – ought to have damned them utterly and irrevocably in the public mind. But the mere presence of these abominations, the mass produced seven dwarfs in the millions of front gardens of England, show that these rules and precepts were fallacious. Incidentally, the presence of the aforementioned dwarfs shows another thing, namely how easy it is for any awakening of public taste to be driven back into the same old groove of respectability and fashion imposed from above, how easily the search after better things can be withered into a relapse into worse things.
>
> (1943: 9–10)

Mass Observation's back-handed compliment ('the masses' take us by surprise by 'welcoming ... intellectual entertainment') ends with something of a slap, in that gnome enthusiasts are constituted, ultimately, as victims of a mass culture 'imposed from above' (ibid.). Gillo Dorfles, writing in the 1960s, construes mass production in similar terms, reserving special contempt for celebrity and pop culture gnomes (who include another Disney figure, Donald Duck, in addition to JFK and Pope John and are seen as distinct from 'actual' gnomes).

Even to friends of gnomes and their suburban homes such notions of pretentiousness and mass-culture prejudice prevail. Gnome defender Russell keeps up Dorfles' distinctions between the actual and the debased. Lamenting the decline in gnome standards she comments: 'Traditional gnomes were almost nowhere to be found, only an execrable array of crudely produced ornaments' (2005: 5). Paul Oliver *et al.*, advocates of the suburbs and suburban style, rehearse a familiar tendency to construe consumers as unwitting in an otherwise sympathetic attempt to account for gnome popularity:

> their significance, *if unknown to those who use their replicas as garden ornaments*, belongs to the collective unconscious. But their presence in Dunroamin's gardens, while reflecting their timeless symbolic meaning, was by no means an archaism; they performed a function that was more whimsical *though their owners were unlikely to be aware of it.*
>
> (1994: 169, my emphasis)

In addition to embodying key Dunroaminer values (e.g. a well-deserved rest after a hard day's work – gnomes traditionally are miners, woodcutters and farm labourers and enjoy a spot of rational recreation in fishing, gardening and smoking pipes), gnomes are 'phallic symbols' (ibid.: 170). The authors elaborate:

> Their form and, in their pointed caps, their colour, was phallic; the postures of the standing figures priapic. There need be no surprise at the popularity of Walt Disney's Snow White and the Seven Dwarfs among Dunroamin's film-going public in 1938 ... The small pool around which the gnomes sit or recline, and in which they fish in so many gnome gardens, is a vaginal symbol. Together they express territoriality, the determination to settle and populate the earth within the domain of Dunroamin.
>
> (ibid.: 170)

In this semiotic reckoning, nineteenth-century spiritualism gives way to twentieth-century psychologism. Freud and Jung account for the power of gnomes over and above their folkloric, mythic resonances. Gnome lovers, in this portrait are reduced to bearers of a primal structure, unwittingly acting out sexual scenes as they tend their lawns; as 'innocent purchasers' of objects from garden catalogues (such as 'the pair of storks, one erect in masculine pose, the other bent in female submissive posture' (ibid.: 169)). Phallocentrism aside, this account introduces a staple element of comedy into Gnomeville: the standard joke that the respectable classes, despite the restraint and discipline of their best self-presentations, betray themselves as obsessed with sex.

The accepted story of the gnome is, then, bound up with ideas of social pretension, unwitting consumption, myth, folklore and collective psychology. Gnome lovers believe in fairies, are acting out unconscious desires and are in thrall to a voracious pop culture. Gnomeworld, as a result, has been construed as fertile territory for some bad behaviour; or to put it in more anthropological terms, its career biography has impelled new, often violent, episodes in the social life of the gnome.

Designer gnomes

In 1999, in an audacious act of creative destruction, Philippe Starck invites the gnome inside. Napoleon and Attila – gnome tables and stools – transgress the boundaries of the metropolitan hotel lobby:

> St Martin's Lane is a dramatic and daring reinvention of the urban resort. Smart, witty and sophisticated, Philippe Starck's design is a brilliant collision of influences – from the modern to the baroque – that suffuses the hotel with energy, vitality and magic.
>
> (St Martin's Lane Hotel, 2008)

Gnomes are the new black, ingredients in the new *smart* aesthetic. In 2007 the Design Museum in Ghent invites Napoleon and Attila to participate in a 'rather unusual' exhibition entitled Kitsch, Camp or Design which 'asks the visitor a number of questions'. For instance:

> Questions about the confusing phenomenon of kitsch and camp, about the design aspect of Philippe Starck's garden gnome, about the exaggerated ornamentation of the rococo period. We hope that the visitor will find a few answers in the exhibition. Or perhaps he won't.
>
> (Design Museum, 2007)

Designer gnomes (and art gnomes – Lucas's *Gnorman* could be included here) are hence valued for their transgressive quality. Translating the charismatic ideology of both hotel and exhibition into sociology it is clear that the gnomes' presence is required to disrupt aesthetic assumption and to perform an honoured twentieth-century art function, summarised in Matthew Collings' phrase 'kicking arse' (1999: 7). When the gnome appears in places from which he is presumed to be automatically excluded – *mass culture in the gallery? folksiness in the metropolis?* – he extends a conversation that begins with Duchamp's *Fountain* and continues through pop art and the millennial Brit-art antics of the Young British Artists.

Such iconoclasm is rendered literally in a recent discussion of the role of kitsch in 'the new ceramics': a smashed gnome emblematises the consecrated relation to gnome-aesthetics (Wiles, 2007) and kitsch serves as a boundary marker of 'edgy' design. To operate in this way, though, the gnome needs to be *tamed*, if not killed (the smashing of the gnome leaves little room for doubt). Designer kitsch and the designer gnome need careful labelling. *Informed* and contained iconoclasm is the name of the game, as BBC gardener Cleve West's garden design (incorporating a group of gnomes sprayed blue and imprisoned in a cage) demonstrates (West, 2007). In case the reduction of the gnomes' surface to a monochromatic scheme was not enough, the incarceration does the trick: a belt and braces relation to gnome fun secures the performance precisely as a performance.

A necessary step in kitsch/gnome rehab consists in reducing objects to their *form*. Once perceived as form, monochrome gnomes operate as signifiers, as the *idea* of a gnome. The actual gnome is visible only in contour and the effect is one of requisite distance – reducing the aesthetic temperature from hot to cool. Monochromatic gnomes (Napoleon is available in black and in limited edition gold) are hence gnome concepts, which speak of a designer's stance: *look at me playing with the idea of gnomes!* Gnome signifiers become usable materials in an aesthetic palette for those with less 'courage' than the *enfant terrible* designer. With Starck having set the pace, subsequent performances and transgressions are protected through reference to Napoleon and Attila; designer signatures insure the risk. Having passed through the consecrated spaces of the boutique hotel and the art gallery, the gnome notion is readied for domestic consumption.

This is by no means to suggest undifferentiated consumption, however. The ability to read the codes of such performances is unevenly distributed, as Bourdieu reminds us: '[Art] objects are not rare, but the propensity to consume them is ...' (Bourdieu and Darbel, 1991: 37). Gnome irony and gnome conceptualism is a restricted game even for consummate players. Cleve West withholds a compliment for a 'fantastic example' of a kitsch garden 'done well' for fear of 'outing' its owner (West, 2007). Gnomes, it seems, contain powerful default settings. Irony can be mistaken for sincerity, should the inverted commas be overlooked. The satirical tableaux of Gnome Magic, a visitor attraction recommended by the post-ironic tour guide *Bollocks to Alton Towers*, potentially pass us by if we lack instruction in *how* to look. The Teddy Blairs' Picnic, the Mr Universe competition, the uninvestigated murder scene, are seen to offer an antidote to the gnomes' 'twee reputation' (2006: 313). In this respect, history appears as a coordinate in this system of objects – a hint of a dodgy past designed to inject a frisson of bad taste and to inform the present scene. Likewise, the Hannibal Lecteresque gnomes that hang out in Selfridges signify precisely through the residual sense of tweeness that informs their shape as they grimace through their metal masks.

Where there is doubt it is probably safer to airbrush the gnome's presence entirely. A recent colour supplement piece on Friar Park (Pearson, 2008), Sir Frank Crisp's Gothic revival mansion commissioned in 1896 and current residence of George Harrison's widow, denies its gnome past and its controversial status as a hallmark of bad-taste gardening. Even in the context of Victorian folly, Crisp's garden is remarkable: the underground gnome grotto pales beside a 30-foot replica of the Matterhorn constructed from Yorkshire stone, complete with tin goat and topped with a rock from the summit of the Swiss mountain (Mowl, 2007).

Friar Park, and Crisp's approach to gardening, occasioned a taste battle (referred to in the horticultural world as 'the Crispian row') in 1914. Reginald Farrer, alpine grower and plantsman, wrote in excoriating prose a barely disguised portrait of Crisp's rockery as part of a contribution to his friend Edward Bowles' edited collection *My Garden in Spring*. Crisp responded by publishing a pamphlet, which was distributed at the Chelsea Flower Show that year and reprinted in *Gardening Illustrated* (Buchan, 2000). The row catalysed a debate over the role of artifice in the garden and the appropriate use of ornament, which resounds through contemporary horticultural conversations. The ban on placing the gnome outside the gates at Chelsea has a long history.

The restoration of Friar Park is cast within orthodox 'lost garden' discourse; Crisp's 'wit and eccentricity' (Pearson, 2008: 47) feature centre stage, alongside his Victorian confidence. Gone is any reference to *bad* taste, 'the Crispian Row' or the underground gnome grotto. Instead, a wistful commentary delineating a 'magic' landscape governed by 'Gothic Fantasy', accompanied by the restoration story – Olivia and George's comparably confident ownership of the mansion and its grounds. The period since George's death in 2001 is marked by further boldness: 'Olivia continues to *move it forward* ... She has also started to *edit* the trees on the boundaries to reveal the views beyond the boundaries' (ibid.: 48). Navigating the space of Friar Park and the territory of good taste necessitates an engineer's stance; careful negotiations of atmospheric conditions: an underground river flows into a shimmering grotto; staged spaces of darkness and light orchestrate an auratic horticultural experience.

Friar Park thus exhibits heterotopic elements, particularly in relation to Foucault's third principle, which details a space 'capable of juxtaposing in a single real space several spaces, several sites that are in themselves incompatible' (1986: 25). Mountains, caves, rivers and pools, Switzerland in England, combine to produce a 'totality of the world'. The idea of this particular garden as constituting a 'sort of happy, universalising heterotopia' (ibid.), a quality that Foucault grants to the garden in general, is compromised by the erasure of its gnomes, however. In disappearing the gnome grotto, Friar Park exercises a form of horticultural hygiene that qualifies it as a *heterotopia of compensation*:

> Their role is to create a space that is other, another real space, as perfect, as meticulous, as well arranged as ours is messy, ill-constructed and jumbled.
>
> (ibid.: 27)

The governing principle of such a space, Foucault speculates, is colonial; heterotopias of compensation operate as empire in microcosm, spaces of civilised order, maintained by bold acts of purification.

The gardens of gnome-haters, likewise, operate through the logic of compensatory space. Further, there are dividends for keeping gnomes out of the garden: TV makeover artists and property developers Colin McAllister and Justin Ryan speculate that gnomes threaten to strip £15,000 from the selling price of the average home (Ward, 2004). Similarly, the actions of garden and other designers who like to play with gnomes – sneaking them into Chelsea or smashing or caging them and spraying them blue – are at best engineering third-principle heterotopic space. Considered in the light of Bourdieu's ideas (1993) around the formulation of symbolic value, third-principle spaces can conjure capital through acts of social magic (here a belief in designer magic trumps one in folklore or fairies). Designers' wit, charismatic ideology for the deployment of powerful taste actions, brings together hitherto unimagined elements and juxtapositions, heavily freighted with the requisite information.

For Liu, such smart aesthetics are utterly complicit with post-industrial, information-age creativities: boutique hotels designed for 'boutique lifestyles' (Harvey, 2008: 32). In the era of advanced creative destruction, the actual materials of the

Figure 12.1 Lifford Lane gnomes

smart aesthetic are subsumed to the manner of their manipulation. Aesthetic land-grab is the name of the game and what was once not cool contains the capacity, in the right hands, to be the epitome of cool. Kitsch, and gnomes as archetypal kitsch, thus offers perfect ground for smart aesthetes and witty designers able to play with the information that constitutes the ideology of the aesthetic. Cool gnomes, therefore, come as no surprise in the age of total innovation.

Gnome retirement

For all those willing to take aesthetic risks and to play the gnome game there are plenty more who see gnomes as beyond the pale; lifestyle TV has been blamed for the demise of the gnome in the British garden (BBC News, 2004). However, if gnomes have become an endangered species, then Lifford Lane Household Waste and Recycling facility has made a name for itself as a place where they can live again. The entrance to the plant has been transformed into a retirement home for not only gnomes but also ducks, lions, otters, frogs, squirrels, crocodiles, swallows, Victorian children, retired couples, owls, rabbits, badgers, cherubs, cats, dogs and toadstools. This is by no means preservation for preservation's sake or a static, *museal* exhibition. The gnomes, together with their companion species (Haraway, 2003), lead rich and varied lives: they celebrate Christmas and support England in the World Cup (three old lions sit in formation on a bench for the duration of the tournament, while many of the gnomes are kitted out with England badges) (Figure 12.1). 'It's always changing', explained one of the chief gnome gardeners, 'if you come back in a few weeks it'll all have moved round' (Interview, 2008).

As an episode in the social life of things, Lifford Lane illuminates a relatively overlooked aspect of the consumption process: disposal. If, as Igor Kopytoff (1986) argues persuasively, commoditisation is a *process* then likewise, disposal is revealed not simply to constitute the end of an object's life (see also Gregson *et al.*, 2007). If, to begin with, the gnomes arrived at the site to be discarded, more recently they come to enjoy retirement. During the course of one of my interviews a gnome arrived to join the community, its owner having consciously separated her waste into the constituent Lifford Lane categories: garden, paper, metal and gnomes. What began as worker creativity has since become something of a community project, as people *donate* rather than *get rid of* their gnomes and other garden ornaments. That said, the gnome garden has always been envisaged as a community project; when the question is asked who it was for – Lifford Lane staff or users of the facility – the answer is clear: 'both ... People bring their families. It gives the kids something to look at while their parents are dumping their rubbish. Everyone loves it' (Interview, 2008).

For Latour, such conviviality would come as no surprise, given his belief in the propensity of things, material objects, to 'construct ... social order'. He continues:

> They are not 'reflecting' it, as if the 'reflected' society existed somewhere else and was made of some other stuff. They are in large part the stuff out of which socialness is made.
>
> (1999: 113)

Lifford Lane gnomes constitute the stuff of both staff *and* public relations: workers, in creating this new life for the discarded, have a laugh with each other and the facilities' users, who in turn promise to find something extraordinary in what usually presents itself as a chore. On a broader plane, Kopytoff's identification of an 'analogy between the way societies construct individuals and the way they construct things' is also illuminating:

> [The biography of the individual] who does not fit the given niches is either singularized into a special identity – which is sacred or dangerous, and often both – or is simply cast out ... Any thing that does not fit the categories is clearly anomalous and it is taken out of normal circulation, to be either sacralized or isolated or cast out. What one glimpses through the biographies of both people and things ... is, above all, the social system and the collective understanding on which it rests.
>
> (1986: 89)

The governing social system at Lifford Lane is tolerant and welcoming, heterotopic in the best sense: 'an effectively enacted utopia' where inclusive community relations are materialised. The damaged and remaindered find a place to spend their final days and are allowed to die of natural rather than 'taste' causes. Some get blown over by the wind and smashed, others weather and 'die' of old age – news reports of a regular cull of the population (BBC News 2004) are revealed by staff to be exaggerated and the result of media hype: 'the management just wanted us to tidy things up a bit'. Seen in the context of makeover mania, Lifford Lane becomes a *sanctuary* for all things remaindered by the prevailing winds of

Figure 12.2 Lifford Lane couple

creative destruction. The obsolete and the outmoded are given an eleventh-hour reprieve, allowed to survive capricious fashion judgements and to retire.

Lifford Lane is thus a heterotopia of crisis and deviation: a retirement home constitutes the borderline between the two spaces for Foucault. In providing a space for the remnants of the creative destructions of makeover artists, staff offer a critical creativity which, to recall Liu, acts as a brake on the 'eternal now'. With this, the garden communicates a message about recycling and, simultaneously, the matter work, the *materiality*, of makeover logic; what certain consumers figure as obsolescence, as waste, is here turned into creative material. More, this material can be figured as constituting a version of Liu's idea of a 'dark history'. To reiterate, in the face of a co-opted creativity founded in total innovation – iconoclasm, transgression, the shock of the new – the job of a critical art is to provide a history 'not of things created … but of things destroyed in the name of creation' (2004: 8). As gnomes gather at the very point where they are expected to meet their end they can be seen to act as material witnesses to the actions of creative destroyers. Creative destruction is, thus, confronted with its material fallout at Lifford Lane (Figure 12.2).

Conclusion

The values materialised at Lifford Lane stand in stark contrast to those of the design world. Creative destruction, rooted in the capital interests at stake in the cool gnome, meets its other in the critical, communitarian creativity at a waste facility. For Harvey, micro-considerations of lifestyle and aesthetics extend into the broader terrain of sociality and community:

> The question of what kind of city we want cannot be divorced from that of what kind of social ties, relationships to nature, lifestyles, technologies and aesthetic values we desire.
>
> (2008: 23)

What happens to gnomes – whether they are loved, tolerated or smashed – can, therefore be seen to speak to questions of social ethics. Foucault's take on the garden – and by extension, the gnome or gnome-free garden – is also helpful: 'the garden is the smallest parcel of the world and then it is the totality of the world' (1986: 25), in that it points to the social visions that are at stake in heterotopology, whether aesthetically accommodating or hygienic. Equally, Liu's exacting interrogation of the nature of creativity in the 'age of advanced creative destruction' allows a distinction to be made between a capital-accumulating, self-interested act and a considerably more expansive, life-affirming art. Further, this distinction shows that the taste game has become more complex and certainly has not withered away. There is, irrefutably, a politics to gnome world, which is all the more powerful for remaining undetected.

Acknowledgement

Huge thanks to all at Lifford Lane Household Recycling Centre (Veolia E.S. Birmingham Ltd) for their hospitality and, more, for their creative inspiration. This chapter is dedicated to gnome lovers everywhere.

13 Christmas light displays and the creative production of spaces of generosity

Tim Edensor and Steve Millington

Introduction

On 30 December 2007 the Bishop of Norwich proclaimed in a Christmas sermon attended by the Queen at her Sandringham estate that:

> Some people, I have noticed around here, turn their houses into minor ecological disaster zones.
>
> (Reported by BBC News, 30 December 2007)

The Bishop was referring to the practice of adorning the outside of houses with animated and multi-coloured Christmas lights, a decorative tradition associated with North America, but one which has become a popular festive custom across Britain. The Bishop's comment reflects a cynicism, particularly within media commentary, towards domestic Christmas light displays, which are denigrated not only on the grounds of their environmental impact, but also because of their aesthetic impact on the built environment. The Bishop, however, was left somewhat embarrassed, as he had failed to notice that amongst the group of 'some people' was the Queen herself, who at that very time had sponsored a Christmas festival, using electric lights borrowed from Blackpool to festoon her Sandringham estate with illumination for the benefit of the local community.

In a context of growing concerns about a declining sense of community, doubts about multiculturalism, and the threats to civil order posed by neighbourhood conflict, against a backdrop of global financial meltdown, a discussion about Christmas lights may appear somewhat trivial. Our empirical research with 22 households in Manchester and Sheffield, however, suggests that displays of household Christmas lights signify new forms of community association, solidarity and neighbourly interaction, in contradiction to negative media representations of the practice.

Accordingly, we argue that Christmas light displays have the potential to transform everyday life in a particular and spectacular way, demonstrating how seemingly banal or vernacular creative practices can invest meanings in, and engender a sense of belonging to, unspectacular, ordinary, mundane places. These displays are a form of creativity that has emerged in communities that largely ignore orthodoxies about design and fashion to produce economies of generosity and a sense of conviviality. The production of such values directly challenges the moral and disciplining imperatives of expert-driven models of urban design, planning and regeneration. Further, they produce new and particular modes of sociality and community bonds, eschewing popular criticism that the festive season has become so enmeshed with consumerism that the values enshrined in the 'spirit of Christmas' have been supplanted by an empty, ritualistic obligation to participate in an anxious and calculated exchange of material products.

This chapter challenges this dominant view of Christmas by arguing that light displays are actually becoming important drivers of social cohesion within marginal communities and neighbourhoods. First, we discuss Christmas lights as a form of domestic or vernacular creativity and briefly examine the conflicts that have arisen around their aesthetic content and impact. Second, we examine the social value of Christmas light displays, to explore the value of the practice to the displayers themselves and to the neighbourhoods in which they are located, exploring how the displays might contribute to an economy of generosity that might produce socially cohesive neighbourhood identities in ways hitherto unimagined in mainstream social policy.

Christmas lights, aesthetics and creativity

The annual ritual of decorating homes with outdoor Christmas lights is an established practice in Western culture. In Australia, for example, Winchester and Rofe (2005) trace domestic electric light displays back to 1947, but they argue that the practice is rooted in a much longer Central and Northern European tradition, whereby candles were used prior to the widespread availability of electricity. Popular cultural understandings of Christmas, however, firmly locate large-scale domestic Christmas light displays within the excesses of American consumer culture, evident in critical accounts about the commercialisation of the festive period (Belk, 1987; Miller, 1993).

The growth of Christmas light displays in Britain, however, does not necessarily represent the homogenisation of festive practices based on an American model of consumption. Rather, our research demonstrates the domestication of the practice in terms of content, class position and geography. In the USA, for example, extravagant illuminations often appears to be associated with large, middle-class family homes, but in Britain they have been largely adopted by working-class households, located within what are considered to be marginal neighbourhoods and suburbs.

We suggest a particular geography of Christmas illumination is emerging within Britain, based on distinctions between neighbourhoods that adopt minimal displays of cool colours such as blues and whites, in contradistinction to the colourful and extravagant red, green and gold lights prevalent in other areas. Displays in these latter districts also tend to feature an array of festive motifs, mixing icons of popular culture with traditional Christmas iconography. These objects and materials act as aesthetic markers, inscribing new meanings and identities on the domestic designscape, which, we argue elsewhere (Edensor and Millington, 2009), also stimulates negative discourses that contend that the displayers are abject 'others' characterised by excessive behaviour and lack of restraint.

On the contrary, we argue that the practice problematises privileged notions of creativity and its instrumental use within various social fields. Osbourne (2003) demonstrates how, during the late twentieth century, professional groups such as psychologists, educationalists and management gurus captured notions of creativity as the basis for personal, organisational and social transformation. In contemporary Britain, the state has also begun ubiquitously to deploy creativity within policies embracing business development, education, employment, regeneration and social cohesion. This is necessary, it is argued, because creative people embody the appropriate skills, enterprise, innovation, flexibility and competencies to compete in a fast-paced, knowledge-driven global economy. Indeed, Scott (2008) refers to the 'cognitive cultural economy', in which the social relations of creativity constitute a new phase of capitalist development and are essential to drive the economic resurgence of cities in the twenty-first century.

Within this context urban theorists such as Florida (2002) and Landry (2000) have exerted a tremendous influence on a new generation of policy makers and regeneration practitioners, who are among the latest groups of professionals to call upon the power of creativity, this time as a solution to the perceived decline of urban cultural vitality. Florida in particular has contributed to the establishment of a doctrine of nurturing creative industries through the formation of centrally located bohemian cultural clusters which comprise members of a *creative class*. Such strategies aim to connect the transformative power of creativity to produce more dynamic, entrepreneurial and even cosmopolitan places.

In the introduction to this book we have identified the critical backlash to this rather uncritical dogma, where Peck (2005), for example, demonstrates how the instrumental deployment of creativity in urban policy is imbued with particular moral imperatives to foster individual creativity as part of a neoliberal agenda. But if there is a *creative class*, then what about the *uncreative class*? In a world of creativity, the state of being 'uncreative' (whatever that means) is implicitly redefined as a problem for the state to deal with and is subsumed into broader discourses which problematise social groups who are deemed to lack the necessary creative skills, cultural tastes and competencies to enter a circumscribed definition of the cultural or knowledge economy:

> Groups that score less well on 'creativity' indexes are deemed to be 'lacking' in particular characteristics; places that are insufficiently attractive to this group are deemed to be 'failing'.
>
> (Oakley, 2006: 270)

According to this logic, the successful inclusion of marginal social groups in mainstream society will depend upon their willingness to engage in a set of practices and subscribe to the values promoted by the elite and expert systems which define the nature of the creative economy. The purported existence of a creative class brings into question the location of creativity, and more importantly, who gets to define what creativity is.

Allen Scott (2008), drawing on the work of Csikszentmihalyi (1997), argues that creativity is socially situated and bounded, moulded by existing knowledges, skills, habits and modes of socialisation, and locked into particular modes of expression, or what Scott refers to as 'aesthetic and semiotic archetypes'. For Scott, the creative class is an empty signifier, a meaningless concept, because it is abstracted from specific social conditions. The challenge, therefore, is to unpick dominant discourses of creativity and aesthetics, and develop greater critical engagement through empirical studies which attempt to ground the analysis of creativity, and account for its transformative potential, within embedded local experience.

We contend that Christmas light displays provide an example of embedded practice and that displayers can be considered creative. Our research interviews in Manchester and Sheffield reveal how displayers operate within their own intimate and context-specific value system, marked by a refusal to engage with popular and 'expert' notions of good taste or design, but in a way that does not denigrate the values of other groups in the community either. Rather, our informants displayed a blissful ignorance and indifference to critical media debates about Christmas lights and established notions of good design. When questioned about how they decided to arrange their displays informants typically responded:

> Well it's all higgledy piggledly. There is no style about it – they go up the same way every year.
>
> (Nell)

> I just throw it up ... it is that there, that goes there and that will do. It depends what I pick up ... I just fling it up: 'Will it fit there? Well, fine.' I have never thought about it. No I have never actually thought about it. It is organised chaos. That's the key.
>
> (Debbie)

> Some of them have got lights but most of them don't even have the decorations up and I think that's dead miserable. *It's Christmas!* ... Christmas to me is not a religious thing it's more of a family thing. It is about when you take time off from the trials and tribulations of life and just take time off to appreciate what you have at home.
>
> (Yvette)

With reference to the creative act, Barry (1999: 4) argues: 'what is *inventive* is not the novelty of artefacts in themselves, but the novelty of the arrangements with other activities and entities with which artefacts are situated'. In this sense, then, Christmas light displayers are indicative of an approach that is practical and habitual but does not reflexively define the creative process – they just do it! Their displays are part of a wider set of vernacular creative practices which challenge dominant ideas about aesthetics and creativity and provide examples through which established doctrines that link creativity to economic priorities can be critiqued. This connects to the idea that creativity cannot be defined in advance;

it is experimental, always in a state of becoming (Hallam and Ingold, 2007). In contrast, formal attempts to link creativity to social and urban transformation may reduce creativity to a set of moral imperatives regarding the production and ownership of cultural commodities, channelling appropriately creative people and activities into urban-renewal scripts of bohemian quarters, creative industries and class. Osbourne asserts: 'you're not a creator if you just sit on the line, identifying yourself as such' (2003: 512).

However, in a context in which home decoration, gardening and ideas about 'good taste' increasingly occupy a dominant area of the media in 'lifestyle' programmes which advise on esteemed aesthetics and 'good' design (Bell and Hollows, 2005), neighbourhoods in which Christmas light displays are prevalent have generated considerable negative criticism, in which they are dismissed as brazen and tacky, excessive, an affront to religious values, as a source of neighbourhood conflict, and even as a contributor to global warming. Our research, for example, revealed the following comments about Christmas lights, representative of numerous similar responses:

> Tacky? They are environmental vandalism on the grossest scale. If people want garish decorations, put them where others don't have to see them.
>
> (Brian, UK BBC website)

> I love Christmas and enjoy setting up the tree and decorations with my family but find decorations of this type tacky and offensive. It is perfectly acceptable to decorate the interior of your house to whatever level you choose but to go to these extremes on the exterior has nothing to do with celebrating Christmas but is simply a case of one upmanship with the tacky neighbours.
>
> (Paul, North Yorkshire BBC website)

Colourful, chaotic and excessive displays of green and gold lights and animated displays seem to attract most criticism, treated as unacceptable public transgressions. Implicit in comments like those above is an assertion of how the domestic realm *should* be organised to produce respectable landscapes – that is, in neighbourhoods which uphold particular class-based norms, tastes and aesthetic conventions. There is, it is argued, a time and place to display appropriately measured feelings of merriment and festivity, without transgressing established social convention. However, mediated representations of households displaying Christmas lights also extend into the wider construction of class difference in Britain:

> Sooo Chav-like. Funniest thing is that the majority of these houses with big light displays are on council estates where the majority are on benefits. A sense of wasting money (my taxes no doubt) and wasting electric (not very environmentally friendly). People need a reality check and instead of wasting my money maybe find a job and some taste.
>
> (*Manchester Evening News* website)

I nearly crashed my car when I saw this blinged up scummy chav house on an estate near me in Middlesbrough – known locally as 'middles-b-ROUGH'. I bet they are spending all of their benefits on leccy tokens! Scum! Best bit is it's probably all nicked.

(Chavscum website)

Stanhope is a nice picturesque village in Weardale, Co. Durham. Nowhere is safe from being ruined by the filthy charver scum however. The picture doesn't do it justice as the reindeers & santas are all animated. Please God burn it down.

(Chavscum website)

These emotional public comments perhaps reveal less about the displayers of Christmas lights themselves, and more about the people making the accusations. Contestations about Christmas displays are unequivocally grounded in contemporary discursive practices in Britain which seek to reaffirm social distinctions through the denigration of others within a hierarchical ranking of people, places and objects in alignment with dominant class-based power relations, mobilised around discourses of excess and restraint. Ideas about what constitutes 'good taste' therefore become inscribed on landscapes, in this case through the deployment of artificial electric light, forming a basis for the assertion of class identity and contestation. In this way, landscapes of class are being reconstituted and take new forms of expression in space.

Such comments also share the fear that messy and excessive displays denote messy and abject individuals. In many other social and cultural arenas emotional and affectual responses of disgust, revulsion and fear are mobilised and repeated within representations which vilify, satirise or denigrate working-class cultural practices and bodies (Haylett, 2003; Lawler, 2005; Skeggs, 2004; and Tyler, 2008). Contentions over Christmas lights, therefore, connect strongly to broader attempts to negatively stereotype aspects of working-class life, from medical discourses of healthy eating through to the mocking satire and denigration of working-class values, language and fashion contained in the pages of tabloid newspapers such as the *Daily Mail* and on TV shows such as *Little Britain, The Catherine Tate Show* or the pernicious *Jeremy Kyle Show*.

It is difficult to ignore the connections between these class-based assertions about 'good' taste and the valorisation of a particular group of people commonly identified as the creative class, whose aesthetics and activities are valorised above those of others. And with the creative class becoming such a powerful idea within urban policy, the use of culture as a resource through which to achieve social inclusion is highly problematic. For example, Zukin (2006) accuses urban policy makers in the USA of *urban amnesia*, the tendency to write out embedded class-based forms of cultural heritage, when it comes to using culture as the basis of regeneration or image strategies, particularly with respect to challenges over the representation of working-class heritage and identity in the regeneration of former industrial cities. For Bell and Jayne (2006), a central problem within expert-driven initiatives is their resolute failure to engage working-class communities in visions for cultural regeneration. Arts and cultural regeneration, it would seem, grinds to a halt when it encounters the grittiness of working-class lives and districts.

Contemporary urban social policy needs the kind of imagination that can understand something of the texture of poverty and working-class lives as ordinary and extraordinary ways of being. Without such thinking, working-class people and places can only ever be 'less than' those in whose image they are reconstructed (Gibson and Klocker, 2005: 101).

Both of these class-based articulations through which creativity is evaluated highlight how a plethora of creative practices outside the realm of the creative class can be ignored or impugned. And while the widening access to tools of creative practice and the growth of domestic creativity afforded through new technologies and communication networks is further increasing the potential for new creative practices to emerge across a range of sites and locales, as Markusen (2006) argues, most cultural policies fail to acknowledge everyday spaces of creative practice, such as dispersed spaces of production and performance, decentralised neighbourhoods, small towns or rural landscapes, and, we argue, working-class housing estates.

Through its fascination with creative classes and trendy bohemian enclaves, culture-led regeneration circumscribes the meaning, value and location of creativity. We are not concerned with denigrating those who have been identified as members of the creative class, but rather seek to extend understandings of creativity beyond the doctrine of arts and cultural regeneration, to investigate alternative values associated with creativity and to examine more extensively its relationship to space. An embracement of what is considered frankly abject, marginal, un-sexy and non-glamorous opens up a range of potentialities and possibilities for enhancing the lives of people in ordinary and everyday spaces.

The value derived from Christmas light displays

Our attention here is drawn to vernacular forms of creative endeavour amongst alternative and marginal groups, as well as those cultural producers whose creative intentions are to produce non-economic outcomes. We are particularly concerned with what is being created through embedded cultural practices, in order to move beyond a narrow focus on taste or the aesthetics of material culture, to recognise how creativity may produce social collaboration and communication, and value to producers and consumers. This is something that cannot be measured simply in economic terms – for example, creativity may provide the basis for civic unity or a sense of self-esteem, and perhaps even, as Jean Burgess (2006) maintains, a source of dignity for ordinary people. So what do people derive from the display of Christmas lights?

> Personally, obviously we got our little boy but it brightens our lives just to know that we've created something – hopefully, happiness.
>
> (Michelle)

The desire to put up displays each year was unequivocally marked by a sense of generosity. The light displays themselves were not exclusively for the consumption of the household itself but were also designed to please the community at large, to engender and create the 'spirit of Christmas' for local people. Further, the displayers undertook this practice without expectation of any return or contractual arrangement, other than a sense of gaining pleasure from seeing other people enjoy their illuminations. Indeed, the only times money was exchanged were when passers-by dropped loose change into charity buckets that several displayers had placed outside their properties, as we discuss below. The sacrificial nature of giving is perhaps rooted within moral imperatives informed by traditional British, twentieth-century working-class cultural values. This was expressed through the creation of idealistic notions of family, community spirit and neighbourliness at Christmas time.

Too often, critical narratives of Christmas focus on secularisation and commercialisation and ignore the emotional and affectual qualities produced through festive rituals. There are many cynical accounts of Christmas in which popular icons, such as Santa Claus, are reduced to signifiers of materialism, and the whole ritual of gift-giving represents the worst excesses of consumer culture, as if gift-giving is only about giving to receive and involving individuals in anxious choices about what to buy and how much to give (see Belk, 1987). Golby and Purdue, however, maintain that, despite commercialisation, Christmas rituals continue to centre on long-established themes of 'nostalgia, home, family and children' (1986: 83). The desire to give something to those less fortunate or to create a sense of excitement or specialness around the festive period, often motivated by memories of festive childhood experiences, resonates strongly within the communities we examined.

Above all, it is the central importance of Christmas as a festival which informs the desire to have a light display, with Christmas marked out as a particular time in which celebration, sharing, giving and friendliness should be prioritised. However, the displayers did not reveal any particular religious fervour, contrasting with studies in the USA, where religious themes continue to occupy a more central position in the annual ritual (Lacher *et al.*, 1995). Rather, the exuberant nature of British displays was connected more to a sense of the carnivalesque, frivolousness and fun in which Christmas is conceived of as a time for excess, and therefore should be celebrated in style.

In most cases the initial motivation for exhibiting Christmas lights was to produce a sense of family togetherness and pleasure, particularly to provide delight to child family members, a common-sense notion that may be generated by a sense that in the past, such pleasures were unavailable.

> [We never decorated] because I had a shite childhood ... [and now] ... I want to make sure the kids remember the Christmases as happy times. I suppose you could say to some degree I'm compensating for that, but I just love it – it's a family time innit?
>
> (Stuart)

Yet while this initial impetus may be family oriented, it quickly extends into a desire to engender wider pleasure, amongst friends and neighbours, particularly their children:

Christmas is about sharing and this is a way of sharing and doing something for the community, for people other than myself. It was just for my own kids and then for everyone else's kids around. They start to expect it now. When it gets to November they start saying, 'When are your lights going up?' I get a lot of enjoyment out of it. When I see the faces go past the window with a big smile on their face, not just the kids but adults.

(Trevor)

This is further articulated through the emphasis that Christmas is a time that is concerned with conviviality, communal pleasure, neighbourliness and generosity, in contrast to the rest of the non-festive year (Lacher *et al.*, 1995; Pollay, 1987). In some cases, the engagement in communal and neighbourly participation goes beyond the responsibility for festive cheer being provided by one or two key households and develops to encompass most of the houses in a street or close. While these closely connected households may also participate in communal events at other times of the year, such as Bonfire Night and Easter, Christmas is the most important time for such celebrations:

Everyone in the street gathers round; it's a big event ... It turns into a party. One year we had all the old people from the old folks' home come down with their zimmer frames. My wife is going to give them all a bottle of whiskey next year when they make it down ... The other neighbours do it too – so it's a street display ... You see we are a very friendly street. In the summer we have barbeques and wash the cars, every year we organise a camping trip with all the families together ... We used to have bonfires every year, but after 10 years that got boring, so we get the kids all dressed up for VE day. On the turn of the millennium in 1999, we started collecting money from every house. We got £2,000 together. We bought an army tent ... We put it up on New Year's Eve. We all did our own street acts ... We all put our displays up together, go out together, put the lights up as a street and then on the 6th we all take them down. The week after the first Sunday when the lights go up we all have Christmas dinner, the adults only. We go out and book somewhere and have our meal.

(Steve)

Finally, the expansion of notions about the distinctive time of Christmas and the need to think of others is reinforced by those who are also concerned with creating displays for charitable reasons as well as for neighbourly purposes. Several displayers have small charity boxes, which encourage onlookers to make monetary donations for particular charitable causes.

Three years ago I got throat cancer, so I decided to do it for Christies [local Cancer Hospital]. I just put it all in the collection box for Christies, so they don't know I'm doing it for them. But we also do it for the kids. There is a little kid at the end of the road, and all the parents bring them through here to see the lights. And nothing ever gets pinched and that, because I think people appreciate what we've done. I look forward to people talking about it. They say, 'When are you putting them up?' and I say, 'When I've got a bit of time'.

(Roy)

Such acts clearly run counter to negative media representations of Christmas light displays and the people and neighbourhoods responsible for producing them. Illumination is commonly perceived within contemporary debates about the commercialisation of Christmas as symbolic of excessive gift-giving, a lack of restraint which fuels market exchanges and produces an even greater distance between secular and traditional understandings of the festive season. For critics, Christmas light displays are not gifts to the community. Rather, they represent greed, selfishness and a desperate desire to mark status, and further serve to heighten the growing obligation to buy and exchange Christmas gifts. However, we argue that it is inappropriate to extrapolate this logic and apply it to all cultural contexts, since such critiques are often argued from a particular middle-class social position, reflecting anxiety and repugnance towards gift-giving. It is difficult for such critics to comprehend why displayers go to so much effort and cost to produce elaborate illuminations without obvious material return. Notions of gift-giving within socially and economically marginalised social groups and neighbourhoods, we argue, remain poorly understood by critics and social commentators.

Dwelling on the excesses surrounding the exchange of material goods as exemplary of the ways in which Christmas has become an empty secular practice that maintains the norm of market exchange relations, obscures, for example, the case of Roy. There is an underlying altruism to Roy's gesture, a sense of concern for others that was shared among all the Christmas light displayers we interviewed. Bromberg (this volume) asks us to consider the failings of totalising narratives, such as Marxian accounts, that reduce all human interactions to commodifiable exchange relations. Shifting the focus away from material practices and exchanges, beyond the economic outputs of creativity and contractual relations, and by examining how the creation and exchange of non-economic values produces emotive and affectual responses, provides us with a glimpse into an unexplored economy of generosity.

Christmas lights may well be assemblages of material products and commodities, and the power that drives them ultimately has to be purchased from a privatised utility provider, but this is not what is being exchanged within the communities we examined. Rather, the lights are vehicles for the exchange of immaterial qualities of wonder for children and adults, charity and festivity, and are a mechanism for community bonding, for thinking of others less well off than oneself – in short, creating the 'spirit of Christmas' replete with its traditional understandings intact, in a way that eludes constraining discourses of restraint and aesthetic reflexivity, as the following displayer captures:

[They bring us] pleasure for usselves and for other people I think. As long as you bring a bit of happiness into someone else's life put them up, blow your electricity bill. You can always pay that when it comes in. Neighbours enjoy it and people knock on the door and congratulate us for bringing a bit of cheer to the road, which is something and nothing really. It's nothing to think about really to put them up as long as it benefits other people, it benefits us as we put them up. Other people going by, the dancing snowman, they love those as they go past and they sing.

(Joseph)

For many communities, the lights had become ritual features of the landscape. And for many householders, despite the often wearying effort required to install the illuminations, they had become a communal obligation that neighbours expected and anticipated. Nell articulated the comments of many displayers when she declared:

It was funny this year as we were late putting them up and the neighbours were coming and saying, 'Is there anything wrong Nell, are the lights not going up?' I said, 'They will go up we are just busy at the minute'. You must keep your audience happy!

(Nell)

Moreover, most of our informants had assembled their displays over many years, often at minimal cost to themselves, for instance by typically purchasing cheap lights from budget retailers in January sales for use the following Christmas. Sometimes the lights had been handed down from one generation to the next. Either way, it was understood within the local communities that such displays were not excessive splurges of consumerist excess to assert distinction; rather, they were the result of years of budgeting and gradual accumulation, representing genuine and heartfelt attempts to make Christmas a time of enjoyment. They were collected through measured decisions, planned for, anticipated and exacted in a rigid temporal framework, as the following quotes illustrate:

Some of them cost a fortune, but you see the crafty thing is that you wait till Boxing Day or the day after, then all the prices go down. That snow globe cost me £58, but they're in the catalogues for £99. You see, so you have got to look around and shop around … you have to grab a bargain when you see it. If you see it on offer, whatever time of year, you have to grab it.

(John)

Every year we've strove to buy one more. We have bought one for this year and we bought it when we bought the tree, in the January sales. It's the helicopter right on the wall. That's the new one for this Christmas.

(Mr Juett)

Furthermore, there is little evidence from our research interviews of any aggressive competition or vandalism, no sense of one-upmanship, out-doing the neighbours with bigger, brighter lights, a practice often cited in media accounts. Indeed there was very little obligation or expectation of neighbours to join in. If there was competition, then it was nothing more than a friendly rivalry, in good spirits.

Conclusions

Given its ubiquitous nature, domestic creativity uniquely possesses power to transform everyday space and the lives of ordinary people. Christmas lights do this in very particular and spectacular ways, demonstrating how creative practices can invest meanings into everyday space which begin to transcend associations with banality, alienation or subjectification. The displays reveal how a sense of belonging can be produced in mundane domestic and communal settings through creative practice.

This examination of Christmas lights also suggests how we might better understand the spatialities of creativity and move away from the constraints exercised on creativity by the pull of the centralised cluster and other valorised sites and subjects, and signifies, together with many other vernacular creative practices, how creativity is constituted by those outside the metropolitan centre.

Most importantly, the displays have emerged in communities that largely ignore orthodoxies about design and fashion, to produce economies of generosity and conviviality. The production of such values directly challenges the moral and disciplining imperatives of expert-driven models of arts and cultural regeneration, interrogating claims about who and what is or is not creative. When examined in context, such displays contribute to notions of identity and belonging within particular communities, as a source of bringing families together, neighbourly interaction, charity, solidarity and memorial.

The effect of the glow of Christmas lights in the communities we observed suggests that particular affordances between the public and the private realms had been created, to reveal a practice that Silk (1998) might describe as 'caring at a distance'. In putting up their displays, individual householders were clearly thinking or feeling about others, and not just friends or family, but also passing strangers or simply the community at large, often individuals that the householders did not know intimately. Christmas may have been a unifying concept within the practice of light displays, but the motivation to display was not predicated on a fervent religious moral obligation to spread Christian values within the community. Rather, the householders shared a sentiment that Christmas was a time of thinking of others, of inclusion, and ultimately about sharing festivity and fun.

We conclude, therefore, that the Christmas light displayers are moral actors within an economy of generosity. Bromberg (this volume) defines the economy of generosity as practices which encourage 'sharing and non-competitive forms of sociality', in contradistinction to the 'gift economy', in which 'reciprocal exchange may be tied to feelings of obligation'. Unlike the gift relations described by Mauss (1967), there was no implicit reciprocity assumed by the displayers. In fact it was widely recognised that they might not receive anything at all for their efforts, which locates the practice firmly outside of Derrida's (1992) 'ethics of symmetrical reciprocity' (see Barnes and Land, 2007). At the same time, the practice did not invoke some moral crusade about community spirit. Clearly, the displayers got something out of it themselves, a sense of satisfaction in giving pleasure to others. As Barnes and Land suggest, this 'ethical conduct is best exemplified by practices of sacrifice, devotion, and love; practices which might be the stuff of ordinary, everyday life' (2007: 1072).

However, as Raffel (2001: 113) points out, 'generosity along with any other form of moral behaviour must in some sense be a product of communally produced and identified standards'. Although the displayers were clearly thinking and acting in a generous way, their actions were also imbued with notions of social justice and sharing. Their generosity was not 'natural' or spontaneous. Nor were the displayers mischievously playing with the rules of capitalism through everyday resistance. Instead, their actions were bounded and constrained by particular ethical principles and obligations regarding their social position. Without doubt the householders believed the pleasures and sentiments of Christmas ought to be available to all, regardless of background and status, revealing a deeply held communitarian spirit, a sense of solidarity and shared identity, or what Barnett and Land (2007) might describe as a form of 'weak univeralism', which connects the practice of Christmas displays to deeply set working-class values, for example, the notion of giving something up for others, making sacrifices or 'going without' so that others might benefit, values which today seem to belong to a bygone age of social democracy and working-class solidarity.

The sometimes vicious media critiques that focus upon these displays seem to indicate certain individualistic and neoliberal values, often oriented around attempts to achieve distinction mobilised by claims of aesthetic worth and tastefulness. The evidence we present above clearly shows how misrepresented Christmas light displayers are in Britain, with notions that people are motivated by sentiments other than a self-centred, egotistical consumerism, seemingly beyond the realms of imagination. Perhaps, then, Christmas light displays are an all-too-visible symbol of wider societal anxiety concerning generosity and attitudes to others or encounters with strangers, whereby a surfeit of generosity is to be feared and critiqued within broader discourses of excess and restraint.

Acknowledgement

Thanks to Emily Falconer, Panni Poh Yoke Loh and the research participants.

Part IV

Vernacular creativity and everyday life

14 Challenge, change, and space in vernacular cultural practice

Ann Markusen

Introduction

Around the world, high cultures, packaged cultural products, and vernacular cultures exist cheek by jowl, especially in larger cities. Where colonization, industrialization, immigration, deindustrialization, and urban residential dispersion have disrupted prior settlement patterns, the variety of cultural expressions and practices can be profuse and highly differentiated. Space and place foster or suppress particular traditions, as communities, organizations, and individual artists use cultural expression to preserve and pass on tradition, encourage young people and new arrivals, solve problems, mobilize politically for change, and bridge across cultures. Within vernacular cultural communities, tensions and challenges complicate how artists and communities organize and sustain cultural activities, especially in rapidly changing environments.

Vernacular cultural practices encompass a wide range of activities that are distinguished by their expression of community values and their inclusion of many participants, in contrast to the individualized and professionalized creation or reproduction of art or culture by experts detached from a community frame of reference. In Gross's brilliant distinction (1995: 16), they value most highly the inherently abundant abilities in people rather than the inherently scarce (e.g. virtuoso violin playing). They can also be defined as what they aren't – part of the elite canon of high culture or fine art in their respective societies (e.g. European classical music, Japanese Noh theatre) or conceived, produced, and distributed by commercial mass media (TV, films), though as Jean Burgess argues elsewhere in this volume, elements of vernacular cultures are sometimes taken up in these other two spheres. Although such practices often endure or emerge among groups who are marginalized, vernacular cultural practices need not be the province of the oppressed only. Members of different layers of social strata in urban and rural communities around the world maintain, modify, and create new expressive, participatory cultural activities.

Vernacular cultural practices carry with them symbolic functions that integrate and maintain social reality for their participants (Gross, 1995). They are governed by conventions, in the same way that high art and commercial music and films are (Becker, 1982), but such conventions are less protected by institutions such as universities, funders, and large, well-heeled arts-producing and -presenting organizations that nurture the canon. Groups both inside and outside seek to define vernacular conventions, preserve them in the face of mass commercial culture, and alter them in ways that will adapt them to changing community circumstances. Furthermore, certain groups within the community may develop material stakes in particular versions of vernacular culture and its presentation. These include proprietors of cultural space, and artists who may wish to earn income through their artwork. Vernacular culture may also be divided internally, by class and age especially, with respect to aspirations for practice and innovation.

In capitalist society, artistic and cultural practices are valued for their economic muscle, so that those, like high art and mass commercial productions, that assemble large budgets, whether of earned or contributed incomes, are considered the most successful. For instance, Americans for the Arts (2002), the major arts advocacy organization in the US, concentrates on demonstrating the economic impact of arts organizations, arts businesses, and artists as its central lobbying theme in the quest for public funding. In contrast, vernacular cultural creativity is often organized by individuals and groups who ask little or no economic return, since they are gifting their work to the community or engaging in it as a shared expressive practice. These artists and activities are thus not included in most definitions of the cultural economy that rely on statistics based on occupation or industry (Markusen *et al.*, 2008). Yet their cultural work is of comparable if not greater significance. Indicators probing the value of cultural practices in communities have been developed by Jackson, Kabwasa-Green, and Herranz (2006).

Space plays an important part in the fostering of and challenges to vernacular culture. Often, vernacular cultures must rely on borrowed spaces for participation and presentation – churches, plazas, community centers, for-profit restaurants, or casinos. If the ownership of and access to these spaces are contested or if the neighborhoods or towns hosting them are changing rapidly, it becomes more difficult to rely on particular spaces and to decorate and stage them in ways that enhance cultural experience. However, many groups have been able to build, claim, and manage space to stabilize and invigorate their own cultures.

This chapter explores how space, region, insider/outsider challenges, and complex, dynamic community social structures shape vernacular practice. I begin by emphasizing the centrality of challenge and change, illustrated with a historic case of Native American visual art in the US Southwest. I then show, using two California-based Asian-American dance groups, how challenges within the community are accommodated through innovations in traditional art forms. I then explore the creation and role of dedicated space as a way of rooting and providing continuity for vernacular cultures while fueling vitality and innovation, using five cases that differ in their origins, spatial and mission orientation, and organizational and governance structures. I conclude with some tentative comparative inferences and identify important routes for future research.

The challenge of change in vernacular cultural practice

Most vernacular practices are rooted in evolved or intentional communities. The character of each community changes over time, and its substantive practices may change as well in response to internal innovation or changes in material circumstances or location. On the one hand, vernacular cultures buffer communities from disruptive, often violent, change, confirming identity and guiding their responses. Yet in a communicatively interpenetrated world, communities of practice are continually exposed to new ideas, art forms, and cultural movements from which they select and reshape their own practices. In addition, entirely new communities of cultural practice may emerge in response to demographic, economic, technological, and/or political disruptions. Some of these themes have been explored in excellent in-depth studies by sociologists and anthropologists (Alvarez, 2005; Jackson, Kabwasa-Green, and Herranz, 2006; Peterson, 1996; Wali, Severson, and Longoni, 2002).

Communities often respond to external challenges to their identities and livelihoods by intensifying and defending their cultural practices, on the one hand, and altering them, out of necessity or a desire to embrace new ideas and opportunities, on the other. Internal challenges also alter vernacular practices, when one group (sometimes in response to external pressures) challenges the leadership or content of received cultural practices. In this section I explore each of these conceptually, illustrating with case studies.

External challenges

Vernacular cultural practices can be challenged in a directly physical, spatial fashion by incursions of others into one's territory. Such a process characterizes the ongoing challenge to Native American culture and survival in North America on a broad continental scale, but also what happens in many urban neighborhoods when an outsider group, higher-income whites or an immigrant group, moves in. External challenges can also be non-spatial, such as the penetration of external cultural ideas and forms through television, films, and the internet.

Communities respond to these challenges in a number of ways. First, they may invest greater energy in shoring up, codifying, teaching, and performing cultural practices to ensure their survival, availability, and visibility for their own members and prohibiting access to or practice of external cultural modes. This may involve the investing of new roles in specialists whose job it is to preserve and enliven cultural forms. There is also the possibility of atrophy in these traditions, as the power of external cultural practices divides the community. Second, if a community is poor and small in size, there is the potential for its cultural creativity to become marketable, even 'exotic' goods and performances for outsiders. A community's artists may participate in such markets, but often negotiate their own terms on compensation and on content, including the withholding of sacred practices from the market nexus. Some communities' cultural practices become the targets of acquisitive outsiders who wish to collect and possess their artifacts and performances and who offer financial patronage in return, often causing traumatic disagreements within the community.

We can explore combinations of these challenges and responses with an overview of the long and complex history of southwestern US Native American interaction with white culture from the mid-nineteenth through the twentieth centuries

(Markusen, Rendon, and Martinez, 2008). Before Spanish and American intrusions and displacement, Native cultural practices were intertwined with material life, in the making of pottery, jewelry, and weavings that had functional as well as decorative and expressive purposes. But as their land holdings shrank and their exposure to Western tools increased, the Pueblos and Navajos began exchanging their pottery, jewelry, and weavings with white traders, in return for manufactured goods and staples. This involved, over time, alterations in their work, as traders reflected affluent easterners' demand for Navajo rugs by favoring particular styles and introducing vivid commercial dyes from Pennsylvania (Brody, 1976; Kent, 1976; Wade, 1974; Webster, 1996).

Some Native artists moved out of their communities and beyond an exchange relationship into paid work for commercial firms. The Santa Fe railroad and its partner, the Fred Harvey Company, hired potters and weavers to create their works in front of travelers who could then buy the pots and rugs in the hotel's gift shop. Historians debate the extent of exploitation in terms of wages, control over working conditions, and integrity of Indian work (Deitch, 1989; Dilworth, 1996; Moore, 2001). When cars began to displace trains, many Indian potters began to market their work directly to tourists who drove directly to the pueblos (Deitch, 1989).

National cultural elites also challenged Native cultural forms and practices, forcing negotiations that charted a new path for visual artists. In the early twentieth century, American anthropologists and elites such as the Rockefellers believed that Indians were disappearing peoples and that their culture and artwork would die with them. In New York and Santa Fe museums, they began to collect and preserve traditional Indian work. A white prairie woman, Dorothy Dunn, set up the Santa Fe school for Indian artists to teach Natives how to paint in the traditional way: flat, pictographic figures untainted by Euro-American innovations in light, shadow, and perspective. While Dunn has been criticized for suppressing artistic creativity, her initiative provided many Native artists a chance to develop their talents, move on to art colleges, and re-discover and use traditional materials and pigments (Berlo and Phillips, 1998; Bouton, 2007). Eventually, some of her own students became her greatest critics and began to pioneer new forms of Indian artistic expression. Yet they had been acculturated into painting and sculpture, not predominant Native forms, and their sense of purpose for their work had been distanced from a community-embedded practice.

By the 1950s, American arts elites shifted their focus away from preserving artifacts and wished to patronize contemporary Native artists, but insisted that the latter leave their traditional ties behind and become fully modern. In a seminal conference in 1957, the Rockefellers and other patrons proposed to create a school for US Indian artists in Santa Fe, the Institute for American Indian Art (IAIA), but only if students fully embraced modernism (Anthes, 2006: Gritton, 2000). However, the Institute's first Director, Cherokee textile artist Lloyd Kiva New, allowed Indian artists to work in any genre with any subject matter (Anthes, 2006). Some used traditional materials (turquoise, leather, clay, beads) to depict modern Native themes, while others employed modern media (paint, print, metal, and wood sculpture) to explore traditional Native myths. This broad-minded IAIA ethic nourished generations of Native American artists, including many who became well known and most of whom returned to their own regions and communities to work. The fact that management of the Institute remained in Native hands and that only Native teachers were employed made such a stance possible.

By the 1970s, Pueblo, Navajo, Hopi, and other southwestern Indians had developed robust markets for their pottery, weaving, and jewelry, much of it controlled by artists who remain embedded in their communities and urban neighborhoods. Contemporary visual artists compete in fine art and art fairs with work that reflects Native themes and is diverse in technique and materials. Casinos commission work from well-known Native American artists or their own tribal members, further diversifying sources of income and opportunities for large-scale work.

This evolutionary path was long and tortuous, with many setbacks. It took place within a larger genocidal context where American Indian policy tried to dispossess Natives of their land and sovereignty and forcibly separated children from their families in boarding schools that forbade speaking their own languages and practicing their spiritual beliefs. Elsewhere in the US, Native American artists do not enjoy comparable patronage, nor IAIA-type institutional support. This history demonstrates the challenges to vernacular cultural practice when an oppressed community faces powerful economic and political forces beyond its control, including the introduction of new cultural materials and modes of expression by outsiders who aggressively attempt to alter their cultural practices through force, displacement, or commercial or patronage relationships.

Internal challenges in Asian-American dance

Despite the tendency to celebrate community and to infer it from the spatial concentration of distinctive practices, most communities struggle with internal divisions over content and control of cultural practice. Traditionalists have a stake, sometimes material, in the preservation and replication of received cultural practices, demanding purity of form and content and resisting innovations that younger or more worldly members of their communities may espouse. Men may claim privileges over women, or vice versa. Older people may suppress new young art forms, while youth may reject the advice or participation of their elders. More prosperous community members may develop property rights or honorific stakes in cultural practices and restrict participation by their poorer neighbors or those who are not fully members of the community (including multiracial or multi-ethnic people). In this section, I explore such internal challenges using Asian immigrant dance forms in California to explore ways in which these struggles play out among contemporary communities.

West coast US cities like Los Angeles and San Francisco host Asian immigrants – Japanese, Chinese, Filipino, Vietnamese, Pacific Islander, and others – who today reside in extensive, loose-knit auto-based neighborhoods. Many bring their dances and music with them, rebuilding temples, churches, and gathering spaces in the image of home as convening spaces for their communities and as ways of teaching their children about their cultures. As their members become third and fourth generation, practitioners preserve and alter traditional dance forms, reintroducing them to communities who have lost touch and adapting them to contemporary realities.

One challenge to vernacular practice is that members of one's own community have changed with their relocation to new places. Two young Filipino-American would-be choreographers, Joel and Ava Jacinto, encountered traditional Filipino dance for the first time as undergraduates at UCLA and devoted themselves to studying it in villages in the Philippines for a year. Eager to bring what they learned back to Los Angeles, they founded their Filipino dance company, Kayamanan Ng Lahi (Treasures of our People). But they found that they could not easily repli-cate the intimacy of the village setting – their people no longer lived this life. They thus adapted the dance movements to the more formal, distanced American theatre format. In this effort, they viewed their role as tradition bearers rather than as individual artists hoarding cultural resources for profit and self-aggrandize-ment. Today, Kayamanan Ng Lahi performs in the region and around the world. It focuses on creating, sharing, and maintaining cultural resources in the process of community building (Markusen *et al.*, 2006: 14–15).

Another challenge is that communities' traditional values, such as attitudes towards gender roles or towards competition versus cooperation, may be in flux. In the Bay Area, Chinese-American Wilson Mah and his brothers teach and prac-tice the traditional Chinese Lion Dance in their community, begun as a way to preserve their heritage. Lion Dance was a highly competitive and male-only dance form, accompanied by powerful music and dazzling costumes. As Mah and his brothers succeeded in building groups among young people, they were challenged by the desire of girls to participate. They chose to open up Lion Dance to girls, including creating a special all-women group for young mothers. They also decided to play down the heavily competitive mystique of Lion Dance so that it could build solidarity and connection within the broader Bay Area Chinese community (Markusen *et al.*, 2006: 62–3).

Both dance troupes demonstrate the role of artistic leadership in preserving and adapting traditional cultural practices in immigrant communities. Dance, in particular, is not a remunerative art form – unlike visual art, it is not collected, nor does it appreciate in value over time. These dancers/choreographers made compro-mises in their careers. The Mah brothers accept no compensation for their dance work and support themselves with unrelated jobs. Joel Jacinto is the director of a large Filipino social services agency and does his choreography on the side. They would not be tagged as artists in formal data sources. Their prominence in their respective communities is a product of devotion to traditional practices, combined with innovations in participation and form fit for transplanted communities in new environments. Their innovations include adopting a village genre to fit the American stage and an urban immigrant audience, altering the gender definition and ethos of a dance form.

The role of dedicated space for vernacular culture

Artists and community leaders often create dedicated spaces to nurture vernacular practices, address community issues, interpret their cultures for others, and tran-scend static conceptions of their own cultures through innovation. Such spaces may convene cultural participants on the basis of common tradition or interest, and/or they may serve a surrounding neighborhood or district. Some may be cross-cultural. They may be run on a commercial, non-profit, or public basis. Often, they are designed and constructed by artists who use their skills to make the space attractive and engaging while serving artists and communities simultaneously. Many involve non-professional community members in artistic creation. In this section, I use five cases of the construction and operation of vernacular cultural space to explore diverse patterns, including who founded each and for what reasons, who is served, how each is governed and operated, and the spatial relationship between each and its neighborhood and regional constituencies.

Factors influencing the formation and success of dedicated spaces include the following, drawing on inferences from a study of 22 artists' centers in Minnesota and selected case studies from California (Markusen and Johnson, 2006; Markusen *et al.*, 2006). First, there is the design choice between serving a surrounding neighborhood or a more diffused constituency through a broader region. Those spaces that aim at neighborhood constituents are often successful in bringing people into their space, strengthening bonds within the area (sometimes across cultural lines), and preserving traditional cultural capital. However, when the community is poor or not well organized politically, sustainable funding for operations can be a problem. Such spaces are also often challenged by demographic or economic changes in the neighborhood that alter the character of participants. Inner-city churches are an example of the latter. Spaces that aim at a larger, more diffuse audience, sometimes along disciplinary lines (visual arts, literature, theatre) and sometimes along ethnic lines (e.g. an Italian or Chinese cultural center in a large metro), will draw higher-income participants who can afford to commute, but will have looser and perhaps contentious relationships with the immediate community in which they are embedded.

Second, the organizational format will make a difference as well. Some vernacular spaces are run by city governments, as in San Francisco and Los Angeles neighborhood arts centers, or receive budget support for renovation, IT, or physical space upgrades, as in New York City or Minneapolis. The budgeting and regulatory process is often formidable and can result in heartbreaking delays; but in good times, they are spared these concerns. There is also, for publicly funded spaces, the problem of content and freedom of expression – public officials may discourage outrageous, violent or overtly sexual presentations, especially if exposed to children. Non-profit organizations, in order to receive the tax benefits for their operations and their donors, must abide by government regulations regarding structure (boards of directors), accounting, and reporting. Critics argue that, in recent years, non-profit arts organizations have become more like commercial businesses, and some community groups have eschewed non-profit status altogether (Markusen *et al.*, 2006). Small, community-serving non-profits have a very difficult time competing with large, well-heeled 'high arts' organizations for wealthy patrons and foundation funding. For-profit vernacular cultural venues – and these exist, as the following cases show – often devote a portion of their space to sales of goods (CDs, T-shirts, artists' work) and services (performances, food and drink) in order to underwrite their community service, but this leaves them vulnerable to fickle consumers.

Third, vernacular cultural spaces may be operated under very different governance structures. Those taking a public or non-profit form are constrained in this
regard, but in for-profit or unincorporated spaces (three of the five cases), considerably creativity can be employed in governance design. Two of these cases involve
a leadership team that makes all decisions, a highly effective form of governance
but one that relies on benevolent motivations and may pose problems of succession, especially if the principals have a material stake in the space. A third case
involves an extraordinary experiment at democratic governance without a hierarchical leadership structure. And a fourth case, a non-profit, demonstrates how
bottom-up creation of a common cultural space led to an unusually responsive and
member-dominated board structure. After brief profiles of the five spaces, I draw
some tentative inferences regarding relative success and the causes thereof.

Café Royale: space for jazz and African-American history in San Francisco

Jazz, a truly American art form kept outside the canon until recently, remains
chiefly composed, performed, and appreciated in commercial spaces. In an
unusual variant, San Francisco's Café Royale nurtures musicians and visual
artists by providing space for work, rehearsals, and exhibitions as well as performance (Markusen *et al.*, 2006). Programming includes a winter history series in
which big band jazz innovator Marcus Shelby teaches the evolution of jazz in its
African-American context, bridging the conventional gap between audience and
performer. The jazz space, in the heart of the city, serves a city and even regional
market, though its regulars are more apt to come from neighborhoods closer in.

Café Royale opened in 2000, when Shelby's business partner Kate Dumbleton
purchased it, envisioning a performance space with a non-profit feel where musicians, writers, visual and spoken-word artists could come together and feel a stake
in the place. Besides food, drink, and live events, Café Royale offers a large down-
stairs rehearsal and discussion space and two painters' studios that are bartered for
visual arts services. Not just a performance space, it is a place of spontaneous
connections where artists end up working together. Shelby's ensemble performs
here frequently, as do others.

Café Royale is also designed by partners Shelby and Dumbleton to be an
educational center. In 2005, during Black History month, Shelby ran a series
of public talks there, playing important recordings and reviewing the history of
jazz, traveling from the blues to big band and bebop. The gatherings explored the
historical context, including slavery, and succeeded in drawing a repeat crowd
from the broader community.

In part because of the financial stability afforded by the Café, Shelby has been
composing new music on African-American themes. His full-length big band
composition *Port Chicago*, commissioned by the Equal Justice Society, explores
the racism and politics of a 1944 Bay Area naval explosion. Recorded on the Noir
label, it has been performed for diverse audiences throughout the Bay Area as a
way of remembering and healing. Shelby subsequently composed and produced
a jazz oratorio, *Bound for the Promised Land*, about the life of Harriet Tubman.

These compositions are infused with history and politics, filtered through the language of jazz.

Shelby and Dumbleton envision an ongoing mix of composing and presenting, more projects like Port Chicago and Harriet Tubman, and a continuing role bringing together jazz musicians, artists, and their audiences. They ruminate on performing space and how to extend what happens at Café Royale beyond performance per se – how music can be used to tell stories, teach, discover oneself, and how it works in the community. Café Royale is a pioneer cross-disciplinary and cross-cultural space where the normal, Western distance between performer and audience is diminished and where the larger community is invited to explore the embedding of African-American jazz in its larger historical context.

KAOS Network: space for young Los Angeles musicians and filmmakers in Los Angeles

Serving the Leimert Park neighborhood of Los Angeles, the heart of the city's African-American community, KAOS Network is a state-of-the-art multi-media center that offers young musicians and filmmakers a place to do their work (Markusen *et al.*, 2006: 12). It was founded in 1984 by filmmaker Ben Caldwell, an artist and social activist who made a commitment to live in the neighborhood and chose to use cultural expression as a way of creating futures for neighborhood youth. Currently, KAOS's three buildings host a small store, a screening room, a recording studio, and space for open mic hip hop, yoga classes, teaching, and other activities. For more than twenty years, it has provided life-changing opportunities for artists of color.

Caldwell began making films as a master's student at UCLA in the 1970s, a member of the Los Angeles School, a group of politically minded black independent filmmakers. An artistic innovator, he wanted to make films that are more like jazz, more African. His seven films, including *Medea* and *I and I*, are experimental, with influences of magical realism. He has also been using film to document the work of Los Angeles African-American artists, including legendary jazz pianist-composer-community-icon Horace Tapscott and the artists associated with the Watts Towers Arts Center.

KAOS Network opened with a digital arts program for youth, teaching video production, television, and film. In 1992, it added Project Blowed, a Thursday night open mic event. To encourage young artists, Caldwell drew on African and especially jazz formats, but made space for them to develop their own genres and styles, including hip hop as a widely influential art form. From the start, KAOS has been a for-profit operation but one where fees are set low enough to ensure accessibility. Costs are covered by staying active and open throughout the week. KAOS's structures were built slowly, out of pennies, selling clothing and CDs as a part of the enterprise. Caldwell believes that working at the grassroots levels would broaden KAOS's distribution system over time. For instance, the long-running Project Blowed event and recordings from it serve as a platform to reach black audiences globally, while simultaneously engaging youth in the neighborhood.

*El Centro Cultural de Mexico: Santa Ana cultural convening space
for Mexican immigrants*

In the mid 1990s in Orange County's Santa Ana, one of the poorest US mid-sized cities, a group of Mexican immigrant women began meeting and hosting community gatherings designed to preserve and pass on their cultural traditions (Sarmiento, 2006). In 2002, they moved into their first of several Santa Ana spaces, offering music, dance, art, English, literacy, and theatre classes, and space for community participation. They set challenging principles for their group: they are inclusive of all art forms, from 'punks to Jarochicanos' (Sarmiento, 2006: 7), and they operate on a participatory basis, with decisions made in consultation with all rather than by their volunteer leaders. Unusual for immigrant groups, they have chosen to work across borders as well as locally.

El Centro celebrates the culture of the Veracruz region, including Son Jarocho, played on unique stringed and percussive instruments, made in Veracruz, at fandangos, i.e. festivals based on traditional music and dance. El Centro offers dance and music lessons in various styles, taught by volunteers. It hosts older women knitting together and young people experimenting with contemporary American music and spoken verse. It cultivates mutual respect: 'mothers may volunteer at a rock show, and punks may volunteer at a Son Jarocho presentation' (Sarmiento, 2006: 10). Decisions about programming and space are made by large, inter-generational meetings of the whole, a strategy that gives all members of the community a sense of ownership of El Centro and raises the visibility of volunteer teachers and administrators. A shared leadership structure decentralizes responsibilities among five volunteer leaders. The structure is sometimes cumbersome, and makes it impossible to contemplate non-profit status, but so far has yielded important benefits in community support and solidarity.

El Centro nurtures relationships between its community and Veracruz, a poor region that has lost many younger residents to the US. It brings teachers from Veracruz to teach Son Jarocho and sends Santa Ana youth with their innovative versions of traditional music to Mexico. It commissions instruments and costumes from Mexico and markets them on a national network in the US, generating income and economic development in Mexican communities. Transnational exchange of Son Jarocho through El Centro is part of an expanding cultural movement and is not confined to the place of origin of its participants.

El Centro has confronted ongoing challenges, including dismissive attitudes on the part of city government. Santa Ana is a first home for recent Mexican immigrants, especially from Veracruz, and by 2000, Latinos accounted for 76 per cent of the city's population. But the City of Santa Ana has tried to remake itself as a gentrified artist-friendly place, including making way for a University of California Irvine student 'Artists' Village' (housing for arts students that includes studio space) and hanging banners that state 'A Place for Artists' over a newly developed area that is displacing Mexican families. El Centro receives a tiny chunk of community development block grant moneys from the city, covering only 17 per cent of its operating costs, compared with much higher public spending for most other area cultural organizations. El Centro continues to fight gentrification and to stake its claim to downtown turf.

The Textile Center: cross-cultural space for women fiber artists
in Minnesota

Minneapolis's Textile Center was founded early in the 2000s to increase the stature of fiber art and act as a gathering, work, and feedback space (Markusen and Johnson, 2006: 63–6). It serves as a home for weavers, sewers, knitters, and other textile artists in Minnesota, almost entirely women. Drawing members from the broad metro region and surrounding states, it is not a neighborhood center but rather a gendered and cross-cultural place, as Hmong weavers, African-American quilters, and Native American beadworkers are increasingly included in the fold.

The Center is the creation of weaver Margaret Miller and three colleagues from the Minnesota Weavers Guild. In the 1990s they grew weary of supporting themselves in isolation, unable to find presentation space and audiences. Textiles were not considered an art form, disparaged as women's hobby work. Few museums and galleries would buy or exhibit them. The weavers spent a year traveling the region and meeting with anyone who might be interested: quilters, knitters, weavers, basket makers, textile shop owners, rug makers, jewelry makers, wire artists, and beaders. They discovered an extensive underground community of textile artists meeting in church basements, libraries, VFW (Veterans of Foreign Wars) halls and homes, including 1,700 quilters meeting in a machinists' hall.

The physical space that now hosts the Textile Center took seven years of perseverance to find, fund, and renovate. The founders toured hundreds of buildings and considered five floor plans, to find an affordable space to house looms, quilt layout, a fabric print-making lab and messy dye shop, as well as offices, classrooms, and meeting space, a gallery for solo and member shows, a juried gift shop, and the nation's largest textile and fiber art library (books, videos, and magazines). Anyone interested in creating textile and fabric art can belong to the Textile Center for a modest annual fee: membership reached 3,000 by 2005. Artists of all ages and abilities choose from a panoply of classes – from embroidery and lace making to needle arts – that simultaneously offer accomplished textile artists an opportunity to teach and to earn income. Fiber artists may apprentice to masters (including fiber artists from all over the world), see the latter at work, hear how they have built their careers, and get feedback on their own work. Every artist member may display her work in one of the Center's eight gallery shows each year.

The Textile Center is governed by a board that includes one member from each fiber arts organization under its umbrella, as well as accountants, business owners, attorneys, and others with textile experience. Its founders bucked the trend toward non-profit boards composed only of professionals and wealthy individuals, insisting on active governance by women artists themselves. Although it was founded by white women of European-American origin, it has reached out to serve women of color in regional communities who practice and innovate in the fiber arts indigenous to their cultures. As a space dedicated to textile work, the Textile Center is a novel development in vernacular culture, created by women artists as a convening, networking, and service home, building connections among practitioners of a vernacular art form who were previously divided by both ethnicity and distance.

Homewood Studios: space for visual artists in an inner city neighborhood

In high turnover and insecurity-ridden inner city areas, residents have often fought to take over vacant or tax-forfeited space, enlivening it with cultural activity that will reverse neighborhood decline and offer youth an alternative to drugs and gangs (Markusen and Johnson, 2006: 68–71). An example is Homewood Studios, serving Minneapolis's Near North and Willard-Hay's racially, ethnically, economically, and socially diverse neighborhood where single-parent and low-income families predominate and 50 percent of the population is under the age of 18. Homewood Studios, a community-based gallery with six artist studio spaces, nurtures neighborhood artists by providing a place for them to create and show their work. The visible presence of working artists in a neighborhood contributes to the vitality, self-image, and connectedness of that community. It also offers opportunities for young people to use art as a way of tackling community issues. Homewood's transformation from a vacant storefront to a transparent, inviting community space animated by artists and artwork led to the expulsion of drug dealers from adjacent vacant properties and attracted new cultural venues.

Long-time Homewood residents George Roberts, a retired high school English teacher and printmaker who desired his own studio space, and his wife Bev, a community organizer, began in 1997 to create Homewood Studios as a way to combat neighborhood housing stock deterioration and drug use by converting a vacant corner building into an art-focused community center. Rejecting the non-profit route, the Robertses took out a mortgage on their home to finance the purchase, to be repaid through studio rentals. From City tax increment financing moneys dedicated to neighborhood councils, their neighbors voted to help fund the building rehab effort. Over the five years that it took to complete the building, the Robertses identified over a hundred artists in the neighborhood who could either rent space or animate it with exhibitions, projects, and activities.

Homewood's gallery and studio space supports emerging artists, who must be neighborhood residents, and provides them a stepping stone to wider audiences and evolving careers. In its gathering and work spaces, artists reap advice on their work and talk with other artists about career challenges. Homewood Studios also gives them the tools to reach out into the community. In a neighborhood where many young people are not making good choices about education and careers, Homewood uses the arts as a way of helping them to identify and choose among options, including becoming artists themselves.

Homewood Studios has had an important impact on internal and external perceptions of the neighborhood. The gallery shows not only expose the work of neighborhood artists to a much larger audience, but exhibit the positive qualities of the neighborhood to visitors, demonstrating, in the words of one of its artists, that 'violence isn't the only thing here.' As the new, non-drug anchor for the local commercial strip, Homewood has created a safer environment for the community, demonstrating how a long history of turf issues and racial divides can be overcome by providing space for latent cultural practices. Its unique and low-budget, non-hierarchical structure offers a viable model for smaller-scale artistic space in neighborhoods that might otherwise have nothing. It demonstrates that a supportive arts community, for both artists and residents, can thrive in a difficult and balkanized neighborhood while serving as a seedbed for improvements in the immediate built environment, both commercially and residentially.

These cases, all artist-initiated, exhibit striking variations in the design, creation, and operation of vernacular cultural spaces for diverse communities. In each, an individual artist, an artists/non-artists team or groups of artists saw an opportunity to further their own work while serving a broader constituency, devoting considerable amounts of time and ingenuity (and sometimes their own financial resources) to the effort. Not all vernacular spaces are initiated by artists, but most have close partnerships with key cultural actors. Overall, the managers of vernacular arts spaces and the artists they serve emphasize the significance of dedicated space. Cultural organizations that operate out of an office that does not act as a convening space have a difficult time winning loyalty from those they serve and nurturing the myriad connections and shared experiences that are possible when a space offers and harbors spontaneous contacts among cultural workers and community members on a recurring basis. The preciousness of space is especially mentioned by organizations that experience or fear displacement or loss of their quarters.

Although the research underpinning this review of space has not been broad enough to draw definitive inferences about spatial constituencies, organizational form or governance structures, some suggestive conclusions can be offered. Those spaces that are primarily oriented to a neighborhood or local constituency enjoy greater ease in convening participants and encouraging lasting connections among them. This does not prevent them from reaching out more broadly, as KAOS's international broadcasts, Homewood Studio's service to youth and artists in the surrounding metro, and El Centro Cultural's strong ties to Veracruz demonstrate. The more regionally oriented spaces – Café Royale and the Textile Center of those showcased here – face less-certain patronage and even the possibility that competitors in suburban areas may begin to compete with their offerings. More specialized art forms or more dispersed constituencies (e.g. ethnic groups that are scattered around a metro region) may, however, be better served by a regional focus.

The case studies and others in the larger research efforts demonstrate that the organizational differences between non-profit, commercial, and unincorporated formats shape prospects, but no one form is clearly preferable. Each offers trade-offs among forms of financial security and subsidies, flexibility in decision making, responsiveness to constituents, and (in the case of the public sector and vis-à-vis public service missions of non-profits) freedom of artistic content. In the US, the rather unique non-profit form of space provision offers considerable financial advantages in tax forgiveness and deductibility of donations by patrons, but imposes heavy accounting burdens and an often complicated governance structure that diverts leadership attention from service to members or community. Many youthful space creators in recent years, observing the rigidity, huge staff costs and sometimes mediocre results of non-profit cultural groups, have decided to go the for-profit or simply unincorporated route.

With respect to governance, public and non-profit space is subject to constraints on decision making, embedded in political and legal structures. Even with these, there are many degrees of freedom in decision making, as the Textile Center's unusual member group board presence demonstrates. The two spaces run on a commercial basis (Café Royale and KAOS) and a third unincorporated space, Homewood Studios, illustrate the range of flexibility and initiative accorded their founder-managers, all of whom operate as benevolent executives and whose impact is a function of his/her deeply held personal values and commitment to community. They demonstrate that you do not have to become a non-profit to make substantial contributions in multiple ways to a community. Each does, however, face succession problems in the future, which may also involve ownership questions regarding the space itself. Where they are fully unincorporated, cultural groups operating spaces like El Centro Cultural are free to experiment with governance, including participatory decision making.

Conclusion

Vernacular cultural practices are continually challenged and changing. Their development is more diverse than in high culture, since they lack prescribed organizational formats and long-standing public and elite patronage. Artist and community group initiative and leadership have been central to preservation and innovation, as the cases reviewed here have shown. Taken together, the cases show the purposes of practicing and housing vernacular culture: preserve tradition and teach it, build confidence among community members (especially youth and women), address political issues, and bridge cultures and regions.

External challenges include urban and rural demographic change that can vary from genocidal forced migration and dispossession, as in the case of Native Americans, to the depopulation of Appalachia through coal and agricultural decline, to the growing affluence and suburban dispersal of immigrant populations. Internal challenges include tensions over preserving traditional forms versus innovations that often incorporate elements of external cultures or new technologies or materials; how to alter perceived inequalities in rights to participate in one's own cultural practice (along age or gender lines, for instance); whether to serve only the current community or to encompass outsiders; and how to deal with the needs of some artists for income versus the historic practice of culture as a volunteer participatory activity, as gifting.

As vernacular cultures evolve, they confront content and organizational challenges. These tensions deserve greater research attention. One dimension involves how the search for space, funding, and support from public, non-profit, and commercial sectors alters cultural practice. Funders have norms and performance criteria they bring to bear. The struggle over control of twentieth-century Native American visual art demonstrates a long process of encounter and negotiation that altered Native painting and sculpture, disembedding artistic practice from community settings and professionalizing artists, but in ways that offered Indian artists continued freedom to choose media and themes and to address issues of their own community. Does this trajectory also hold for other vernacular practices under pressure?

Second, what happens when vernacular cultural creations become commodities or collectors' material and possessed by museums and other outsiders, as in visual art, or are copyrighted, produced and distributed by others? Nineteenth-century Navajo rug makers altered their patterns to suit the preferences of remote eastern buyers. Ojibwe communities have fought for years for the return of their sacred scrolls from US museums and collectors, finally winning a Congressional law requiring all public institutions (but not private owners) to return such materials (Oakes, 2006). Native historical figures like Pocahontas and Sacajawea have become known through non-Native writing and film, often gross distortions of their people and own lives. There are good critical studies of many instances of such alienation and appropriation, but almost no adequate comparisons across vernacular cultures.

Third, how can vernacular communities deal with tricky issues of artistic freedom and youthful artists' penchant for critique and introduction of external artistic modes when the targets are venerable cultural practices? El Centro Cultural, through its extraordinary consultative process, has developed an explicit ethic of respect for both old and new expressions through face-to-face listening and volunteer time commitments to support of artistic practice across age barriers and varying degrees of assimilation. The Textile Center, due to its inclusive constituency building and member group seats on its board, also resists the development of fiber art hierarchies. Are these exceptions? In contemporary Native communities, struggles over the use of sacred texts, drum music, and artifacts like stone pipes, dreamcatchers, and eagle feathers are often vociferous and judgmental. In many other ethnic art forms – Japanese Butoh dance, for instance – founders and purist advocates reject experimental forms that venture beyond the evolved set forms.

Vernacular culture practices have received very little attention from contemporary cultural policy makers and developers of urban space – El Centro Cultural's struggle for respect from its city council is a powerful example. This is changing in some US cities, like Chicago, Los Angeles, New York, and Minneapolis, mainly because of communities' self-organization and demands for a share of public resources. Research work exploring causality and outcomes has tended to take the form of individual case studies. Comparative work is daunting – time consuming and expensive in terms of resources. The Markusen and Johnson (2006) comparative study of 22 Minnesota artists' centers took two years and $55,000 to complete. 'Studies of studies,' where a researcher reads through many case studies and summarizes the insights, would be an alternative. Unfortunately, many of the studies of vernacular cultural organizations/spaces are quite uncritical, while critical studies of vernacular expression and content do not address material or organizational aspects. Better research would help communities to articulate their needs and place them squarely in the middle of the policy table.

15 The politics of creative performance in public space

Towards a critical geography of Toronto case studies

Heather E. McLean

Introduction

On a sunny September weekend, parks in downtown Toronto's Queen West and Parkdale neighbourhoods became rowdy spaces for people in their mid 20s and early 30s to play tag, capture the flag and red rover. The Time Out/Game On intervention invited 'participants and viewers to celebrate the spirit of the playground in and outside the park, while challenging our notion of playful space and submission to the rules of the game' (Balzer, 2007). Curated by Toronto artists and playwrights, these games were part of a broad range of interventions in the Play/Grounds participatory performance series that were part of the Queen West Art Crawl, a neighbourhood arts festival in Toronto's downtown Queen West neighbourhood. Some other performance interventions in this series included interactive, site-specific plays in the nearby boutique hotel and Toronto artist Jon Sasaki's installation in the local Salvation Army store, where the windows were taped shut and the space was filled with black light. The space was 'black enough for bewilderment, but just enough for your eyes to adjust so you could find your crocheted toaster covers' (Operation Centaur Rodeo, 2007) and the shoppers were given individual flashlights to shop that day. Funded by the local Parkdale Liberty Economic Development Corporation (PLEDC), the Parkdale Business Improvement Areas and Artscape, a non-profit organization that promotes affordable housing for artists as well as 'culture-led regeneration ... stewarding creative communities, and playing a catalytic role in the revitalization of some of Toronto's most creative communities' (Artscape, 2008), the events animated the streets and brought people together in interactive performances that revealed complex layers, histories and narratives about the two neighbourhoods.

These Play/Grounds interventions are one of two performance case studies from Toronto that this chapter investigates to understand these types of active physical engagement and dialogue within urban space, including the artistic and activist potential of urban performance interventions. It also unpacks the case studies to reveal how they are directly and indirectly linked to larger political debates about gentrification and the territorialization of the 'creative class' at a micro, neighbourhood level. I begin with an overview of the political potential of performance interventions theorized in the performance studies field, and the growing interest of some critical geographers in these art forms. These theoretical connections, I argue, are increasingly important for two reasons. First, these interventions provide activists, artists and academics with unique methods for creatively and critically engaging with urban spaces. Second, they are also important to keep an eye on as culture workers and innovative, grassroots notions of culture are currently instrumentalized by policy makers, planners and developers looking for ways to develop distinctive neighbourhoods in order to market neighbourhoods in competitive cities. This chapter connects writing on the 'politics and poetics' (Crang, 2003) of performance to theorists exploring the role of the arts and aesthetics in the gentrification and promotion of spaces for 'creative class' planning goals. I argue that this latter body of work rarely acknowledges everyday spaces of creativity and creative practices, including small-scale performance interventions. My work attempts to bridge this gap and to critically engage with the entanglements of the politics of performance interventions, the power dynamics of 'hipster-urban' identity formation at the neighbourhood scale, and gentrification. Through the use of performance literature, I also reveal how creativity and class are performed, reiterated and, thus, materialized in these gentrifying neighbourhoods. However, I also illuminate how, through performance, these politics are also challenged and disrupted. This chapter contrasts the Play/Grounds interventions in Toronto's rapidly gentrifying Queen Street West neighbourhood with the work of Mammalian Diving Reflex (MDR), a performance company attempting to confront and grapple with the politics of gentrification and the role of children in these processes in downtown neighbourhoods.

Performance and urban space

In the Play/Grounds interventions, local thrift stores and the shoppers moving through these sites became part of a messy interaction of participants and performers. In the parks, dog walkers, screaming kids, and old men in scooters intermingled with young adults and performance animators playing tag.[1] The artists used empty storefronts, vacant lots, and parks as satellite locations for performance interventions that engaged and interacted with passers-by (Artscape, 2008). These types of interactions are celebrated in the Spring 2006 edition of the *Canadian Theatre Review*, dedicated to a celebration of 'site-specific' performance which provides an introduction to the debates about how performance, the politics of 'creativity,' and urban space intersect. The issue showcases a range of artists who have migrated from purpose-built theatres and galleries to what they call 'spaces of creation' (Nanni and Houston, 2006: 9) including loading docks, waterfronts, houses, and factories with plays, interactive installations, and walking tours. All of the examples explored in this journal reveal the intrinsic spatial dynamics of performance and the entangled relationship between performers and spectators in these creative interventions. According to the edition's editors, theatre and gallery spaces are limited because they are defined by various practices of representation, including acting, directing, design, and dramaturgy, that fill the space of the empty stage with a mimetic reflection of the 'real world' (Nanni and Houston, 2006). In contrast to these traditional theatre spaces, they write, site-specific performances that happen in real and everyday spaces provide opportunities for participatory dialogue. They also insist that within these performance spaces, new opportunities to engage with a range of sites and experience and the multiple layers of spaces are uncovered. These interactions illustrate how performance is more of a 'process than a mimetic image' (ibid.: 9) and that the dialogues created by these interventions provide opportunities to create relationships with sites that challenge how we perceive and construct our identities within these spaces.

Performance theorists and cultural geographers write of the 'artistic, political and pedagogical' (Pinder, 2005) potential of interactive performances like Play/Grounds for engaging residents and artists in interaction: these moments connect residents and artists with everyday, in-between spaces that are both personal and political. For example, the black-lit Salvation Army reveals glowing stains on clothes, traces of the networks we are all embedded in. In the black light, the banal nooks and crannies of the store are lit up in eerie ways to reveal rich and complicated stories. In contrast to the 'placeless places' of theatres and galleries (Nanni and Houston, 2006), Play/Grounds fosters engagement with 'heterotopian' – or the Foucauldian idea of 'threshold' or 'in-between' – spaces. According to Foucault (1986), these are spaces rich in contradictions that reveal people's preoccupations and the political complexities of our everyday lives. Rather than fill a gallery space or a stage with a representation of the real world, these practices engage spaces to interact with these various dynamics. Thus, neighbourhood-based interventions and performances provide praxis-oriented approaches for engaging with the neighbourhoods we live in and the stories that surround us; these practices can enliven and politicize urban space and connect critical geographical enquiry with creative activism.

Performance interventions like Play/Grounds are connected to a long line of writers and artists who engaged with the various layers of city spaces in their art and political practices. These include the surrealists and situationists who used inventive methods to juxtapose random, banal, and beautiful spaces, objects, and things in their work. Walter Benjamin (2002), Michel de Certeau (1984) and Henri Lefebvre (1996) also recognized and appreciated how creative, spatial practices blur the boundaries between politics, art, and everyday lives and space. Recently, cultural geographers have also begun to write about the vital role these experimental modes of urban exploration and expressive play in urban politics (Bonnett, 1996; Pinder, 2005; Soja, 1996). David Pinder's writing on the rise of the art practices like urban pyschogeography opens up new opportunities to examine the spatial dynamics of performance and its potential for introducing ways of interacting in cities and 'writing the city.' Referring to the writing of geographers Steve Pile, Nigel Thrift, and Doreen Massey, Pinder sees the potential of practices of urban exploration for 'feeding into, and resonating with, wider current concerns with rethinking cities and urban space' (2005: 390). There is, according to Pinder and the geographers he cites, great potential in these interactions where interventions like walking tours provide ways of sharing stories, activating our emotions, imaginations, and disrupting the repetitive ways in which we use and categorize spaces.

Collaborative spatial practices, some performance theorists also write, are street-level tactics that help us understand how we perform identities through a 'legacy of sedimented acts rather than predetermined or foreclosed structure, essence or fact, whether natural, cultural, or linguistic' (Butler, 1989: 274), providing glimpses into other ways of existing in the city and interacting with each other. While the streets of Parkdale and Queen West are spaces where people live, work, go to school, shop and commute, they can also be transformed into sites where strangers can meet for a neighbourhood-wide game of capture the flag, bringing humour and physicality to everyday sites. Similar to how drag performances can disrupt mainstream notions of gender and sexuality, these interventions disrupt how we script spaces in our everyday performances and reiterations that become naturalized as 'common sense.' Extending these theories about gender and sexuality to space, we can look to urban performance interventions as methods for 'troubling' (Butler, 1989) and re-scripting our ideas about the purpose and meaning of spaces we dwell in and move through.

According to the event's literature, the interventions 'challenge popular notions of public space and ... explore how diverse populations creatively relate with and within the shared spaces of their community' (Artscape, 2008). Opportunities like this are especially important, as urban dwellers experience the privatization of public space that dulls our senses and restricts opportunities to interact with each other (Pinder, 2005). Instead, these types of interventions help us develop more sensuous and active relationships with spaces and the challenge is to contest the everyday ways we reiterate our relationships with each other and our everyday environments.

The uneven terrain of neighbourhood-scale performance

While enthusiastic about the potential of these types of urban encounters, theorists both in performance studies and in geography have also raised the uneven power dynamics that these performance practices can produce and reproduce. As these practices engage residents at a neighbourhood scale, we need to ask who decides how these spaces and the residents who live in them are scripted and what happens to residents who fall outside the performance's social and spatial boundaries? Acknowledging these tensions, Pinder (2005) warns us why neighbourhood walks that encourage 'exploration' should be viewed with caution, as practices 'freighted with politically charged connotations.' He interrogates the tensions that exist in performance when groups of people feel entitled to 'explore' neighbourhoods and spaces as if they are new frontiers. These dynamics are loaded with colonial political metaphors and problematic race and class dynamics (Smith, 1996). As interventions like Play/Grounds provide unique lenses to view city spaces through, it is important to keep asking whose social imaginary is celebrated and performed in these performance interventions. As certain urban identities are valued and performed, how do these concepts materialize in urban spaces, social, political, and economic dynamics?

Analysing these uneven power dynamics in downtown neighbourhoods experiencing gentrification and displacement is especially important. The well-meaning artists and activists in these types of interventions have their own values and social imaginaries that drive their desire to encourage creativity and play in these neighbourhoods. However, as this imaginary is inscribed in actual everyday spaces where people live, shop, sleep, and hang out, many people outside of these networks are left out and, in some cases, diminished in negative stereotypes in the performances. Already, a lack of communication and understanding between many of the artists and lower-income residents has been an ongoing issue in the Queen West and Parkdale neighbourhoods, as the area has experienced rapid gentrification since the mid-1980s (Slater, 2004). The two adjoining neighbourhoods have experienced the familiar story of gentrification as artists were lured by the grittiness and multicultural character of the areas and moved into the older manufacturing buildings (Bain, 2003). Now that the area is known for its quirky and artistic character, developers keen to build loft condominiums, boutique hotels, coffee shops, and bars have quickly followed the newly established artists and countercultural communities.

Queen West has experienced very sudden change, as two new boutique hotels and several new bars – some displacing smaller art galleries – opened along one four-block strip between 2005 and 2008 (McLean, 2006). Advertised on the website of one Queen West boutique hotel, the area is described as 'full of contemporary art galleries, clever boutiques, interesting restaurants and neighbourhood bars, The West Queen West Art + Design District is the center of Independent Culture in Toronto' (Drake Hotel, 2008). The popularity of the area has soared and the neighbourhood has become a key site for the rapid development of the landscape of 'hipsterurbanism,' where 'one block after the next, block after block, downtown strips are so hip and so different, lined with different bars and cafes and art spaces and restaurants and bars and cafes and art spaces and restaurants' (Cowen, 2006).

Sociologist Richard Lloyd refers to this as the landscape of 'neo-bohemian' development. By 'neo-bohemian' he means people who construct themselves as 'bohemian' in opposition to money and accumulating material possessions (Lloyd, 2005). While these networks might define themselves as oppositional, Lloyd illustrates how these artistic lifestyles and countercultural niches are highly valued in the post-Fordist new economy. The popularity of the 'neo-bohemian' character of Queen West grew so rapidly that a neigbourhood-based organization – mostly consisting of arts and culture workers – had to mobilize in response to the Queen West triangle development where several artists' lofts were torn down to be replaced by three 10- to 26-six story towers (Active 18, 2008). Ironically, one of these buildings that displaced the artists' lofts was cheekily named the Bohemian Embassy Condominiums (Bohemian Embassy, 2008). Only a few blocks west of this rapid development, Parkdale is a stigmatized area undergoing a rapid loss of affordable housing units (Slater, 2004) and struggling with tensions between home owners trying to shut down social services and the needs of new immigrants, sex trade workers, methadone clinic users, and various low-income residents.

These power dynamics rear their head in local media articles on the Play/Grounds performance intervention as well as on the Play/Grounds promotional materials and in Toronto-based artists' blogs. While the aim of these interventions was to activate resident participation and to bring non-Parkdale residents into the community, the stereotypes and images perpetuated in the literature surrounding the events and reiterated in the performance events themselves reveal the same uneven power dynamics underlying the neighbourhood gentrification politics. Ironically, the Play/Grounds' curators present an awareness of these tensions, but also use them to lure participants to take part in the interventions as they state in Toronto's entertainment weekly 'we're forcing people to go deep west, under the bridge' (Balzer, 2007). Co-curator Elaine Gaito describes Parkdale as 'a neighbourhood in transition' (ibid.) where these art projects can provide opportunities to view the dynamic layers of the neighbourhood in active ways. 'Hopefully we can shed some light on places people are afraid to traverse,' she says, and these art interventions are 'going to reveal some things that ordinarily aren't visible … I think it'll be beautiful – maybe an abject beauty, but beautiful nonetheless.'

These loaded descriptions of the residents and spaces of Parkdale reveal the contradictions that arise when artists attempting to engage with site-specific work adopt methodological strategies to 'collaborate' with everyday people in urban spaces. Adopting a 'quasi-anthropologist' (Kwon, 2002: 30) gaze to categorize neighbourhoods in ethnographic terms, these artists construct a relationship where they are asserting the power to classify people and spaces, reinforcing power hierarchies. Often outsiders of the communities they are intervening in, these artists have to be aware of the sense of the uneven politics of engaging in local representations to 're-occupy lost cultural spaces and propose counter-memories' (Kwon, 2002: 32). Embedded in the dialogue framing these performances, these uneven dynamics are, therefore, repeated and naturalized in these neighbourhoods. Performance artist Guillermo Gómez-Peña's insights complement Kwon's writing as he warns of art that attempts to examine 'marginal' or 'peripheral' communities, especially when 'the framing will be done by someone who will never understand the drama of the "outsider"' (Gómez-Peña, 2005).

A further look at some of the Play/Grounds interventions illustrates the tensions of using this neigbourhood, fraught with uneven socio-economic dynamics, as a space to encourage pleasure, play, and discovery. For example, the Play/Grounds Mushrooms of Parkdale Tour raises critical questions about the way in which in the neighbourhood is framed and interacted with. The tour consisted of 'diminutive mushroom sculptures placed in the corners and edges of residential windows and storefront displays,' the Play/Grounds flier writes. The artist attempted to capture the transitions that the Parkdale neighbourhood is experiencing in the walking tour: participants were encouraged to follow a trail of mushroom sculptures through the neighbourhood. The artist writes:

> ranging from life-size to miniature in scale, these mushroom facsimiles invite close-up examination of some of the nooks and crannies of this neighbourhood in transition. Parkdale is experiencing cycles of gentrification which pool and ebb in relation to neighbourhood parkettes, addiction treatment facilities, low-income housing, or the presence of 'fixer-uppers.' The Mushrooms of

Parkdale – as both idealised objects and Oroboric emblems of cyclical decay – act as stand-ins for Parkdale itself.

(Balzer, 2007)

Also, a site-specific play that was part of the Play/Grounds interventions reveals further complicated power dynamics. The Gladstone Hotel, a boutique hotel in the Queen West neighbourhood, was home to a site-specific play where participants could, according to the Play/Grounds literature, interact with and view the play in 'public spaces' around the hotel. Co-presented by Artscape and the local business improvement groups, the fictitious play introduces a storyline about the 'hotel's unofficial and recently barred Karaoke Queen ... and two Queen West scenesters out on the prowl' (Artscape, 2008).

These interactive opportunities in the Gladstone Hotel present even more tensions and contradictions. Two developer sisters, well known for their support of Toronto's art scene, bought the historic building in 2000. They then transformed what had been homes for low-income residents, including seniors and people with disabilities, into a boutique hotel with rooms designed by local artists.[2] While the project was critiqued by some activists for displacing the residents and for partially catalysing the rapid gentrification of the neighbourhood, the Gladstone is now an important hub for Toronto's independent music, arts, queer, and progressive political scene (Cohen, 2006). The site-specific performances in the Gladstone, therefore, took place in a space fraught with the politics of gentrification, displacement, and progressive politics: tensions that challenge us to question who would feel comfortable occupying these spaces and engaging in these interactive plays. Also, what kind of social imaginary is reproduced in a hotel with this contradictory history?

A brief analysis of the Play/Grounds media coverage and artist statements reveals the uneven power dynamics embedded in these interactive interventions. Mould and mushrooms are metaphors used to describe the lower-income residents and the spaces they live in, while explorers are encouraged to discover 'abject beauty' in these everyday spaces. Meanwhile, interactive theatre provides participants a glimpse into the lives of 'scenesters' in the Gladstone Hotel. While the interventions provide innovative opportunities for bringing people together in play, Play/Grounds also reproduces a particular social imaginary about the neighbourhood that also materializes in the ways the neighbourhood is being 'revitalized' in rapid development of condominiums and artist-oriented boutique hotels. Through the repetition of these values in these performances, the neighbourhood is normalized as a degenerate place where many of the residents outside of these artistic networks are not active in voicing their identities.

Performance and the politics of the 'creative' city

Questioning the symbolic power of these neighbourhood scale interventions is important because they are linked to broader politics of the role of aesthetics, culture, and creativity in contemporary cities. These social imaginaries, in turn, materialize in social exclusion and gentrification. These street-scale notions of creativity, performed in interventions like Play/Grounds, provide opportunities to engage physically and creatively with spaces, but they also reinforce powerful neighbourhood-level hierarchies as certain stakeholders perform their identities by narrating and categorizing the spaces, and the residents that live in Queen West and Parkdale. Increasingly funded by business improvement groups and boutique hotels, these neighbourhood-scale interventions intersect with the broader politics of competitive, post-Fordist cities where lifestyle, aesthetics, and fostering 'creative' lifestyles are encouraged by a range of actors. Within these partnerships, street-level and participatory creativity is increasingly linked to the political-economic dynamics of post-Fordist cities (Florida, 2002; Peck, 2005; Zukin, 1991). Over the past thirty years, a key goal of urban politicians, planners, and elites has been both to attract mobile capital, the professional class, and jobs, and also to position urban spaces within the competitive spatial division of consumption (Harvey, 1989; Kipfer and Keil, 2002). As David Harvey writes 'above all ... the city has to appear as an innovative, exciting, creative and safe place to live or visit, to play and consume in' (1989: 9).

As cities are increasingly pressured to compete, various policy makers, planners, politicians and academics have turned their attention to urban development concepts that encourage the production of distinctive and marketable social character in order to foster urban revitalization goals (Florida, 2002; Gertler *et al.*, 2002; Markusen and King, 2003). City boosters, including business improvement area (BIA) groups and planners, increasingly view neighbourhoods as spaces to encourage face-to-face interaction and experimentation in festivals and a range of artistic interventions. Neighbourhood-level performance interventions like Play/Grounds, therefore, are viewed as distinctive cultural amenities and good investments (Florida, 2002).

These attempts to encourage participation and interaction in public space in a neoliberal era of funding cuts, however, encourage ideals of play and interaction for certain class and income groups. For example, while Play/Grounds was celebrated in the local arts press for reclaiming public space, Toronto's recreation facilities were pressured with funding cuts. In fact, the same week that residents were encouraged to playfully interact in Parkdale and Queen West spaces, the City of Toronto announced significant budget cuts to various skating rinks, community centres and libraries. According to one Play/Grounds participant and mother of a four-year-old, the kids had to pee in the park's cedar shrubs because the nearby community centre bathroom was completely closed during the events. This disjuncture raises important questions about the politics of performance and play: can we engage in these activities and also raise important questions about social exclusion and dwindling access to well funded public spaces?

The contradictions revealed in the Play/Grounds case study, where interactive performance interventions celebrate certain people and spaces and assist in constructing particular lifestyles while leaving out and, in some cases, diminishing other residents, frustrate performance artists like Gómez-Peña, who witnesses how performance is quickly instrumentalized for commodifying specific notions of identity and space in urban revitalization politics. He writes:

> If our new audiences are more interested in direct stimulation than in content, can we effectively camouflage content as experience ... One thing is clear to me: Artists exploring the tensions between these borders must now be watchful. We can easily get lost in this fun house of virtual mirrors, epistemological inversions, and distorted perceptions – a zone where all desires and fears and imaginary 'content' is just fading memory.
>
> (2005: 52)

The uneven power dynamics in the Play/Grounds performance examples reveals a need to keep asking the questions raised by Gómez-Peña. This is especially important as these animated methods for engaging with space collide with gentrification. However, in his writing on performance and geography, Pinder also reminds us that discourses of the uneven politics of performance interventions are 'heterogenous and contested' and that performance practices provide opportunities for engaging with power dynamics (2005: 388). A critical performance and cultural geography lens provides us with methods for critiquing these practices and the ways they script spaces which are part of the overarching politics of 'creative class' planning and politics. However, performance practices and theories also give us the tools to interact with spaces in ways that disrupt these performed notions of creativity and encourage critical dialogue. As this next cast study will reveal, there are ways of collaborating with, uprooting and redirecting these notions of 'hipster-urban' identity production that are desired by larger-scale creative class planning policies, condominium, bar and boutique hotel developers.

Mammalian Diving Reflex: engaging with 'hipster-urbanism'

One group of artists in Toronto is attempting to critically engage with these performance practices, raise the politics of gentrification, and break down the bounded social imaginary of 'hipster-urban' planning. Mammalian Diving Reflex (MDR), a company started by Toronto writer, actor, playwright, and activist Darren O'Donnell, has also been working with Parkdale and Queen West neighbourhoods and residents in the Duel in the Dale performance interventions. In his recent writing and interventions, O'Donnell attempts to engage in civic engagement and social intervention as an aesthetic practice. While enthusiastic about the potential of these practices for generating dialogue, he is also honest about his frustrations in trying to didactically teach the public through performance. In his book about his social interventions, *Social Acupuncture*, he writes:

> how did I end up so much time believing that culture had some revolutionary potential? What was I thinking ... was it me who dressed like a businessman and went down to the financial district to dance in the streets, convinced that it had power to affect the withered souls there? Was I so arrogant? Did I

join a Situationist International reading group and walk aimlessly through the city scanning my body for how capitalist planning guides my desires? Did I organize 7 am parties on subways to jar the squares out of their stupor and provide them with a glimpse of a truly liberated soul ... what was I thinking? How did my head get so fat?

(2006: 14)

Writing from the perspective of someone who has participated in a wide range of these types of interventions, in his latest work with MDR O'Donnell approaches performance with an awareness of the ephemermality of these practices. He also moves forward into these interventions aware of the connections between arts, culture, gentrification and the neoliberal constraints of cuts to artist-run centres, community spaces, and funding. He realizes that these interventions are partial, fleeting, and always imperfect, but he also believes that they are still politically and pedagogically important. According to O'Donnell, the idea of the artist-run venue with full community engagement has fuelled the fantasies of activist artists, but within this neoliberal moment, these 'visions had to be adjusted and scaled down; the hip alternative cultural community centre, complete with café, remains elusive' (O'Donnell, 2006: 16). Honest about these constraints, O'Donnell attempts to directly engage with the new boutique hotels and incoming residents as well as kids, public schools, and spaces that are left out of the social imaginary of 'hipster-urbanism.'

In *Social Acupuncture* O'Donnell writes that art practitioners need to counter growing neo-bohemian identities 'intentionally practised by a horny capital in places of speedy gentrification like Brooklyn's Williamsburg and Toronto's Queen West' (ibid.: 24) with what he refers to as 'neo-philistinism.' By 'neo-philistinism' he means we should borrow from philistines who are indifferent to artistic values and prefer material value, but we should replace this desire for building economic capital with a desire for building social capital (ibid.). This neo-philistine approach challenges us to understand art that goes beyond aesthetics, fun, creativity, and imagination to 'directly engag[e] with the civic sphere' (ibid.).

For example, MDR'S Duel in the Dale interventions are a recent attempt to practise this 'neo-philistine' approach (Akimbo, 2008). Raising the tensions of the large number of incoming 'hipsters' or 'neo-bohemians' and the dynamics of gentrification in the Queen West and Parkdale neighbourhoods is among the aims of MDR's work. Rather than backing away from these realities in its work, MDR challenges them head on, in theatrical ways, and finds creative potential in these frictions. In May 2008 it curated a series of performance interventions in the Parkdale and Queen West neighbourhoods. MDR introduces the Duel in the Dale on its website with:

In this corner we have: the kids, 647 of them. A massively diverse crew obsessed with Playstation, teddy bears, fairness and running the world. In the other corner: Artsters, predominately white, mostly from other cites, well-educated in the liberal arts, ready and eager to get drunk at gallery openings and always on the look out for exciting but cheap ethnic dining experiences. Who will be the winna in this ultimate battle?!?

(Akimbo, 2008)

These performance interventions included the company's facilitation of six rounds of encounters between students at the school and artists in the neighbourhood, resulting in interactive performances, public talks, public walks, dinner, dancing, and a film. A main goal of MDR is to encourage dialogue between local kids and local residents, especially the population MDR refers to as 'artsters' and businesses. Within these new relationships and networks, these kids are given a space to interact and become part of how they and their neighbourhood are narrativized both in the performances and in broader dialogue. MDR producer Natalie De Vito states: 'I'm probably one of the reasons the area's being gentrified ... there's a huge artist population in this neighbourhood, but they live parallel lives with the kids. No one communicates with each other. We are interested to see how these two different groups would impact each other's practices – it's as much to help us out as to help the kids. It's a symbiotic, rather than parasitic relationship' (Isaacs, 2008).

Some of these interactions included 'Round 7,' the pairing of the elementary school musicians with the local Blocks Recording artists to plan a concert that was held in one of the local boutique hotels. 'Round 9' involved linking the school's art class in the 'The Art Club vs. Mercer Union Gallery' intervention, where the kids viewed and commented on contemporary art. The Shortcuts and Hangouts Walk, one of the Duel in the Dale interventions, connected residents with Parkdale kids in an interactive walk, where a group of grade-eight kids who had collaborated with artists and animators from MDR led 50 participants on an exploration of the alleys, parks, roti, and poutine hotspots. The mainly adult participants had a chance to explore and engage with the spaces that were meaningful to these kids who led the walkers with a rolling cart equipped with an amplifier and microphones. These everyday spaces included the 'heterotopian' spaces of contradictions and contrast that these kids move through, including the fabled 'U.G' – the 24-hour internet café above the adults-only video store where the boys hang out, the libraries, the local places for cheap snacks, and the graffiti and tagging in the alleys. Many of the adult participants walk past these uneventful sites everyday, but with the 12-year-olds' insights, these spaces became animated with their hilarious, sad and frightening stories.

Although this was a small-scale intervention, it prompted meaningful dialogue in this highly contested Toronto neighbourhood where the growing arts community rarely connects with local kids. This personalized walk provided an opportunity for the kids to lead residents through the neighbourhood and introduce a grade-8er's perspective of Parkdale's spaces. The walkers were led through alleys and streets to the local McDonald's restaurant – the place the kids identified as one of their favourite hang-outs. The stories the walkers heard along the way included: how to avoid getting yelled at by the seniors who live in the retirement home across the street; how some of the kids eat lunch with their grannies in the same home; where the local boys play video games; the mean lady who hates kids hanging out in the greasy-spoon diner who 'horked' in a milkshake; memories of learning to read in the local public library; the free condoms in the basket in the library; their thoughts on local graffiti; and their graphic stories about some of the murders that had taken place in the neighbourhood.[3] The walks provided unique insights into neighourhood stories that many locals might never get to hear about. For example, in one interview De Vito states, 'after the walk – teachers were coming up to me saying "I can't believe the kids took you to the U.G." Teachers have been trying to find the U.G. for years. It's called the Underground or the U.G. because it has an adult video section underground. Apparently the kids drop their basketballs down the stairs as an excuse to go down and have a peek around.'

MDR's performance interventions contrast with Play/Grounds because the Duel in the Dale interventions attempted to reveal the playful, gritty, and banal stories about the kids' lives in Parkdale from their perspective and brought these stories into conversations with the local 'artsters.' Rather than being scripted and reiterated in performance as a scary place that needed a light shone on it by artists, the walks provided opportunities for the kids to explain how they lived in and moved through the neighbourhood. Also, MDR's brochures, websites, and the dialogue in the events explicitly point out the politics of gentrification and connect the kids, many of whom are from low-income, new-immigrant families, with this dialogue. Because the walks were based on the perspectives of workshops with the grade-8 kids, the dialogue on Shortcuts and Hangouts did not overtly raise political issues. However, this walk did reveal the 'in-between,' heterotopian spaces of the kids' lives and provided these kids with opportunities to narrate their stories about how they used the neighbourhood's spaces.

The MDR Duel in the Dale examples in Parkdale illustrate how performance can disrupt and redefine identities and provide a new space that presents opportunities to create relationships with space and with different residents. The attempts to connect the kids with the 'hipster-urban' or 'artster' newcomers in the neighbourhood through the Shortcuts and Hangouts walk, the music, and gallery partnerships illuminates how these relationships and spaces are socially produced and continuously contested. For example, the musical collaboration with local indie rock bands, including the highly political band Lal, in the Gladstone Hotel, breaks down the exclusiveness and boundaries. These actions assert the rights of kids in this neighbourhood both in the physical spaces and in shifting social imaginaries.

The Duel in the Dale interventions have come under scrutiny from writers and artists interested in the politics of performance. Some people critiqued MDR for not directly engaging with the politics of poverty, social exclusion, and homelessness in the neighbourhood interactions.[4] Without this dialogue, the performances were critiqued for reproducing the power dynamics that occur when an outsider group intereacts in a marginalized community. Some critics also questioned MDR's collaborative work with children because it placed these kids in situations where they did not have the power and autonomy to voice their opinions and politics. According to these critics, the kids were awkwardly placed in these performance interventions while the MDR animators acted like 'puppet masters.'[5]

Drawing from a range of ethical questions debated in performance art theory and sociology, the tensions raised by these critics do push us to keep questioning the limitations and potential of working with kids in performance interventions and collaborative community development models.

These critiques are part of broader debates about the politics of working with children in performance art and participatory research (Barker and Weller, 2002; Katz, 2004) as theorists attempt to understand children as important agents in everyday politics and how larger scale processes like economic restructuring and gentrification affect kids' daily lives and spatial practices (Katz, 2004). The Duel in the Dale interventions engage with these evolving questions by providing the space for the kids to assert their identities in the neighbourhood. Through performance, their work is important for shaping the social imaginaries of the neighbourhood and spurring interactive dialogue with a range of residents. Expanding this limited field of dialogue in an accessible, playful way is critical in this neighbourhood, as the social boundaries between the gentrifying 'artsters' and the kids are part of the larger web of the politics of exclusion. Public spaces are increasingly affected by budget cuts while privatized spaces in condominium towers, bars, and hotels flourish. These interventions may not provide immediate political change; however, the symbolic act of asserting the rights of the kids in these spaces encourages dialogue, builds relationships, and provides a more complicated view of the neighbourhood's residents than do the categories reproduced in the Play/Grounds interventions. To quote some performance theorists, these interventions provide 'the gap, the rupture, the spacing that unfolds the next moment allowing change to happen' (Pearson and Shanks, 2001: 403) that might not be immediate. Instead, these interactions broaden the social imaginary of the neighbourhood to include the kids' voices.

Conclusion

This chapter reveals the contradictions in the current trend of using culture-led regeneration to steward certain notions of community building in Toronto. Various creative interventions, including site-specific interventions, are highly valued by business improvement area groups, developers, boutique hotel owners, and planners for bringing creativity and innovation to the micro-level of spaces, neighbourhood spaces, and the residents who inhabit them become enfolded in what are often highly uneven power dynamics. Linking performance theory that examines how we perform our identities and values in spaces, this chapter has explored the ways in which some artists animate downtown Toronto spaces, illuminating interactive methods for engaging in spaces that break down the numbing homogeneity of privatized spaces. However, I have also extended these theories to a political economy critique of the current trend to instrumentalize the arts for urban 'revitalization' goals, and revealed how some performances risk the danger of reproducing exclusionary race and class dynamics in neighbourhoods that become framed as marginal, derelict, and in need of improvement. As the above examples reveal, this is especially troubling when public spaces, community centres, and arts programmes face increasing funding cuts. How creativity, fun, public space, and the notions of the public are imagined and performed in these case studies is part of the broader politics of social exclusion and gentrification in downtown neighbourhoods.

Connected to a rich history of theory and art practices that connect everyday life, built form, and politics, performance theory also directs us to ways in which these interventions can disrupt and redirect exclusionary categories and provide us with alternative social imaginaries, dialogue, and interaction. For example, in the Duel in the Dale performance interventions, MDR attempts to bring the voices of Parkdale's kids into the boundaried social imaginary of what they refer to as 'artsters' in the growing 'hipster-urban' landscape of Parkdale and Queen West. The stories and interactions in these interventions re-narrate and perform the neighbourhood as a complicated place that does not necessarily match up with the notions of fun celebrated in the Play/Grounds games and performances in parks, hotels and thrift stores. Furthermore, the stories revealed in the Duel in the Dale walking tour are based on the realities of the kids, and not on some artist's attempts to 'shine a light' on the abject beauty and places people are afraid to go to in Parkdale. The production of this social imaginary of creativity is increasingly important to make sense of, as neighbourhood-level aesthetics and artistic life-styles are valued in culture-led revitalization. As this chapter has illustrated, these heterogeneous and malleable practices are rich sites of contradiction that can be disrupted by residents, artists, activists, and academics collaborating with people and spaces. These sites offer opportunities to perform more inclusive notions of identities that materialize and shape everyone's 'right to the city' (Lefebvre, 2003).

Notes

1   Personal research of Play/Grounds events, Fall 2007.
2   The documentary *Last Call at the Gladstone Hotel* by Neil Graham and Derrek Roemer follows the Zeidler sisters transforming the Gladstone Hotel from a single-room occupancy hotel to one of Toronto's key boutique hotels.
3   Interviews with MDR's Darren O'Donnell and Natalie DeVito.
4   Several debates about the Duel in the Dale among artists and curators took place on this blog: www.digitalmediatree.com/sallymckay/comment/44812/.
5   www.digitalmediatree.com/sallymckay/comment/44812/.

16 Creativity unbound

Cultivating the generative power of non-economic neighborhood spaces

Ava Bromberg

Society acquires new arts, and loses old instincts.

Ralph Waldo Emerson

This chapter explores the specific accomplishments and limitations of Mess Hall (MH), an experimental cultural center I co-founded with seven others in a Chicago storefront in 2003. MH is a globally networked, locally rooted space sustained by an economy of generosity and conviviality. Occupying a rent-free 680-foot storefront in the far north-side neighborhood of Rogers Park, MH functions as a non-economic space for exchanging ideas and skills, eating with others, developing projects, and celebrating interests with strangers. It offers a free way to socialize in a dense urban setting without being a consumer, an opportunity to build social relations that are at once rare and socially vital. In an urban development and neoliberal policy paradigm that fails to foster communal exchange or engagement that is not pay-to-play, MH makes a particularly poignant incision as a de-commodified storefront in a commoditized landscape. As a *possibility space* fostering creative engagement with the world in which we live and act, I will argue that the MH example stands to expand the discourse on creativity and the city.

While a small and particular case, MH points to possibilities far beyond its walls. It suggests a role for non-economic neighborhood spaces – somewhere between work, home, and commercial life – wherein new social bonds, forms of politics, and ventures of all sorts can be cultivated. As I will argue, developing such *possibility spaces* for deep, unexpected, or provisional encounters in the city is a challenge we cannot neglect wherever we seek, with sincerity, to cultivate creativity and extend that opportunity broadly.

After briefly presenting the context in which it emerged, I will ground my discussion of MH in literatures on generosity and gift economies. Thereafter, I will define some key terms and present the case in more detail, reflecting on lessons from its organizing structure and suggesting directions for future experimentation. My analysis of MH draws from my engagement over the past, five years first as a co-founder and active keyholder – holding a full set of keys to the space – through 2005, and as a participant-observer thereafter. I also draw from writings by and extended conversations with other keyholders who together shape and reshape MH's philosophical and practical operations.

Making space for encounter in the neoliberal city

To contextualize MH's emergence as a non-economic 'alternative,' it is important to note some crucial features of the policy context in which it emerged, namely, how neoliberalism, as a 'non-regulatory, regulatory regime' (Purcell, 2002) has squeezed out non-commodified land uses while systematically shrinking the state's obligation and ability to provide publicly accessible spaces of encounter. It is useful to draw out a few key points about how neoliberal ideology is articulated in cities and their built form. Brenner and Theodore (2002) emphasize that neoliberal ideology applies 'market discipline, competition, and commodification throughout all sectors of society' and has reformulated the role of the state in contextually embedded ways at multiple geographical scales (ibid.: 349). They explore how cities operate as a 'strategic geographical arena' where neoliberal initiatives crystallize into 'geographically uneven, socially regressive, and politically volatile trajectories of institutional/ spatial change' (ibid.).

The doctrine of 'market discipline, competition, and commodification' began to influence policy in the late 1970s and early 1980s in response to the 'declining profitability of traditional mass-production sectors' and the failures of Keynesianism (Brenner and Theodore, 2002: 350). Purcell (2002) pinpoints three major shifts meant to stimulate economic growth during that crisis: 1) the re-scaling of state 'responsibilities and functions up toward the international scale and down toward the regional/local scale,' 2) the move away from a 'policy ensemble of social insurance and social welfare and toward ... workfare, job training, and supply-side intervention in production,' and 3) 'the state (especially the local state) ... increasingly transferring duties outside the formal state structure ... [in] a shift from govern*ment* to govern*ance*' (ibid.: 303).

With less direct support from the federal government, municipalities have become responsible for generating more of their own revenue to balance their budgets. Revenue-generating schemes include a widespread fiscalization of land-use decisions to generate tax revenue wherever possible, even leasing public assets like roads and bridges. In 2004, for example, the City of Chicago gave Cinta-Maquarie a 99-year lease to operate the Chicago Skyway in exchange for $1.8 billion. Private micro-governance districts, from business improvement districts (BIDs) to their residential counterparts (RIDs), exemplify the trend towards using 'market discipline' to fill gaps in municipal services. Under these structures landowners self-impose taxes to pay for hyperlocal services such as trash pick-up or private police forces in their neighborhoods.

Where the infrastructure of public life is concerned, the public libraries, parks, community and recreational centers – crucial spaces open to all residents regardless of their socioeconomic status – often restrict hours and programs if unable to secure sufficient funds from public or private sources. Considering how constrained existing spaces for non-market interactions are under a neoliberal paradigm, I return to a question that frames my discussion of MH: in a landscape saturated by logics of 'market discipline, competition, and commodification,' where do we develop other kinds of social relations beyond our own homes?

It is a rare developer who does not develop every inch of saleable or rentable land on his plot. Where he is required by law to offer public spaces, these are superficially 'public' at best, more likely to be designed and monitored for control than with design grammars that generate meaningful encounters (Smith and Low, 2006; Loukaitou-Sideris and Banerjee, 1998; Larsen and Fowle, 2005). As a result, invaluable sets of social relations are increasingly squeezed into peripheral and private domains. MH emerged in response to this, creating an accessible space for activities and encounters with nowhere else to go, while cultivating an economy of generosity and sparking new collaborations.

Generosity and diverse economies

What are these invaluable relations being squeezed from the landscape? They are, arguably, qualities of life and social interaction with values that cannot be contained by pricing structures. They are not comfortably or adequately accounted for with currency. They appear in the myriad ways we care for our partners, families, and friends, the things we do with or for others without the quantified expectation of payment (either with goods, services, or money) that characterize exchange economies. One hopes the experience of such affective relations is never far removed from our daily existence. Whether 'strong or weak, long-lasting or ephemeral,' Friedmann (2007, emphasis in original) writes, '*non-commodified, caring relations are part of the very substance of life.* They are among the things that make us who we are'. These bonds, Friedmann argues, are not only embedded in the places where we dwell, they *make* places.

The value of caring relations is all the more significant when we consider the context in which we presently live. In early 2009 the ubiquitous and increasingly mobile presence of the World Wide Web – a technology that had non-economic motivating and operating logics from the start (see Benkler, 2006 for an introduction) – has revolutionized the way we connect to ideas, products, and each other. Those with internet access can explore new applications of information technologies that make work easier, news more instantly accessible, and friendship and communities spanning vast physical distances and differences possible. Yet, key to my argument is the idea that the content that makes present and future technologies worth our attention – as well as the connections they nurture – *begins where we do*, in the physical spaces of our embodied lives. In the rhythms of our daily existence we learn how to be together and develop our ethical and aesthetic sensibilities, our integrity, our relationships, our work, the curiosities that drive us, our identities, our selves *with and through our encounters with others*. These formative and generative affective relations need places and spaces in which to grow, not only within each of us or in our homes, but between strangers. Making room for these relations – where we weave the social fabric, encounter and embrace difference in diverse societies – is one kind of vernacular creativity in the spaces of everyday life that deserves cultivation. Yet these qualities are paradoxically uncontainable *and* squeezed by the dominant urban development paradigm.

A brief review of the interdisciplinary literatures on generosity and gift economies further contextualizes the contributions of the MH case. Without analyzing the texts in detail, I will draw out a few major themes relevant to this discussion (see Schrift, 1997; Osteen, 2002 for a comprehensive review of theoretical treatments of the gift as well as Derrida, 1992). Studies of gifts and generosity span diverse fields, and a few strong threads emerge in the scholarship. One body of work analyzes gifts and gift economies as part of larger systems of economic logic, exchange, and calculation. This approach is prevalent in early research on the gift in anthropology and sociology, in which gifts play an important role in developing social and communal relations, creating 'reciprocal obligations' that establish and reinforce social order (Mauss, 1967; see essays in Komter, 1996). Works in moral and political philosophy take a more expansive approach, where gifts and generosity are not necessarily coextensive with, proceeding from, or referring to market exchanges. Diprose (2002) builds on Derrida's and Heidegger's works and discusses generosity as a way of being in the world, of encountering 'the other' in the fullness of difference. For Diprose, generosity is world-forming, an approach to social life and to how one forms one's identity.

In feminist theory, some analysts have distinguished gift and exchange economies as separate, gendered, and competing paradigms, casting the former as feminine and the latter as patriarchal (Vaughan, 1997; 2004; 2007). Others have sought integration, demonstrating how *care*, including 'non-market transactions and unpaid household work' constitute '30–50 per cent of economic activity in both rich and poor countries' (Ironmonger, 1996; in Gibson-Graham, 2008: 615). Going further, Gibson-Graham places care with gifts, generosity, and diverse forms of non-capitalist exchange as a vast network of under-recognized alternative economic activity. In a capitalist or Marxist framework this activity may appear marginal but, Gibson-Graham argues, it is in fact quite central to human communities, embedded in our daily rhythms of survival and meaning, and part of capitalism's vast 'outside' (Gibson-Graham, 1996; 2006; 2008).

The last set of literature I want to introduce involves temporary experiments to create deep public space (defined below), work with surplus, or play with forms generosity and cracks in existing exchange systems. In recent years, this has become an ever broadening domain for creative action. In the most literal sense, there has been a proliferation of artists' projects – with roots in the 1960s – that create spaces of interaction, give away goods and services on street corners and plazas, play with a contentious political issue, or address a variety of topics in publicly accessible ways (see Purves, 2005; Kester, 2005; Thompson and Sholette, 2004 for examples, as well as Temporary Services, 2007, for an investigation into the 'group work' that enables these and other creative efforts). In other arenas, self-initiated 'really really free' enterprises, from free stores and community gardens to forms of open-source production, point to the broadening of a non-economic ethos into many other aspects of work, life, and invention (see Carlsson, 2008 for examples). These pull us further away from a gift economy rooted in exchange systems, either embedded in or running parallel to market economies, and into a generosity economy that creates productive openings, an approach at the heart of the MH experiment.

Definitions

With the above theoretical considerations in mind, I can clarify some terms. The *economy of generosity* at MH encourages sharing and non-competitive forms of sociality. This can be distinguished from the gift economy because it is not tied to feelings of obligation (Bloom, 2006). Rather, it is tied to an idea of *pooling surplus*, that we all have personal surpluses, whether time, money, skills, materials, knowledge, attention, or whatever we bring to the table. When brought together these surpluses are made larger with the potential to cross-pollinate new ideas and practices.

I define *possibility spaces* as constituted by a radical openness to difference – in ideas, life experience, worldview etcetera. In a possibility space one can encounter and be together with others, or bring together distinct ideas, on their own terms. It is an environment where generosity, conviviality, and the *messiness* of coexisting differences, as well as an openness that allows new ideas and forms to take shape in favor of habitual responses or patterns, is evident. Possibility spaces intersect with *deep public spaces*, environments where people can gather in more than a superficial way. They offer participants an access point to participate in *deep democracy* where multiple voices, vocabularies, and differentials of power can 'be together,' debate, and make decisions together. This is necessarily 'messier' than majority rule and up or down votes (see Mindell, 1992).

Possibility spaces are linked to – and in many respects both contain and produce – *radical space*, in that they are sites where dominant relations are likely to be denaturalized (Kohn, 2003). I have chosen a fresh term, however, to capture the emphasis on a space generating new recombinatory forms above and beyond responding to, resisting, or critiquing inherited ones. One can think of a possibility space initially as a rupture, a crack – whether a physical space, or within one's own thinking – where a new cycle takes root. The most important point, philosophically and practically, is that a possibility space is *where*, physically and psychologically, people can instigate new cycles of relations that can in turn shape new kinds of spaces and the physical, mental, and representational worlds those spaces engender.

I also consider *creativity as a public good*. I mean this first and foremost in the economic sense; creativity is a non-excludable, non-rivalrous human resource. In other words, when a person is being creative, they do not prevent another person from being creative. Furthermore, no one can be effectively excluded from participating in creative life, even if in practice many are never overtly connected to it. On the contrary, I consider creativity as an abundant, generative, and pervasive human capacity. In this sense, it can also be viewed as a commons, but one with particular characteristics that, along with free speech, liberal education, and recreation, include the potential for 'infinite returns to scale' (see Rose, 1989). I revisit this idea in the concluding section. For now, it suffices to say it is an all-too-common miscalculation to treat – or fund – creativity as if it were precious or rare, or worse, to forget to nurture it at all.

What is Mess Hall?

Mess Hall is but one of many physical spaces where economies of generosity and conviviality can flourish and ordinary creative explorations are seeded. In what follows, I explore the development of this small case, reflecting on its organizing structure and its capacity to make a space where social, political, and economic possibilities can emerge. While the particular accomplishments of MH over the six years since 2003 are modest, I believe this case presents an opening that can be taken further, and points – theoretically and practically – to potential expansions that support spaces of vernacular creativity in the rhythms of everyday life. MH's relevance to this volume not just as a *place*, but as a political, social, economic, and creative project where others are encouraged to generate and explore their own creativities. It is an incision, a rupture, making room for the development of non-capitalist social relations – and new possibilities – in a monetized landscape.

MH is open for events and hosts everything from potluck brunches and skill-sharing workshops to listening parties, exhibitions, meetings, and lectures by internationally known artists and thinkers. Workshops have covered topics from silk-screening and deejaying to the long-running Sewing Rebellion, a monthly 4-hour drop-in event where participants learn and share techniques for giving new life to old clothing. Hardcore Histories' listening parties pair hardcore music and food from a particular country (Finnish Hardcore; Pancake and Vegan Reindeer Meat Dinner). Lectures bring to attentive audiences work ranging from urban farming to the role of art in social mobilization in former Yugoslavia. 'Free bins' on the sidewalk outside provide for small-scale redistribution, to keep treasures out of the waste stream. Visitors can peruse the small library. There are many ways to participate, with the idea that the event attendee of today may be an event organizer of tomorrow. Global, regional, and local networks converge in the small storefront; some people cross the city, others cross the street, or an ocean to attend events. The space fits about 50 people, but announcements reach over 1,400.

For over six years, MH has occupied one of the eight ground-floor commercial spaces on the 6900 block of Glenwood Avenue, adjacent to the Morse stop on the Red Line 'El' train. The owner of the building, Alan Goldberg, contacted Chicago-based arts group Temporary Services (www.temporaryservices.org) about using the space after reading an article in the January 2003 *New York Times* that featured its work and that of other collaborative arts groups (Cotter, 2003). The members of Temporary Services – Marc Fisher, Brett Bloom, and Salem Collo-Julin – extended Goldberg's invitation to five others (Dan S. Wang, Marianne Fairbanks, Jane Palmer, Samuel Baldwin Gould, Ava Bromberg), who together worked out the original structure of MH. (See www.messhall.org for a list of past and present keyholders, as well as our 'Ten Points' developed in 2007 and our calendar of events.) There are only a few conditions of our lease. MH must participate in the annual two-day Glenwood Avenue Arts Festival, which Goldberg helped to found in 2001. We also have a clause for 'involvement in local and community groups.' We hardly ever pay the token $1 per month rent.

Recent conversations with Goldberg confirm that his support remains a mix of altruism, real-estate interest, and genuine appreciation for the arts. Goldberg's willingness to forgo any immediate profit is a vertical redistribution of wealth in a very tangible form – making storefront space available (Wang, 2004). This act has enabled the value(s) MH supports to have a visible presence and to expand further. The economy of generosity and conviviality started there, and followed with the core 'rule' that we would never charge for events, sell things, 'pass the hat,' or ask for money from attendees. We did not want to create an environment where, even though events are 'free,' people felt pressured to contribute to keep on the lights. Instead we encourage visitors to bring something to share, so there is always something on hand for people to eat or drink. As of February 2009, there were 13 keyholders sharing basic operating costs and coordinating programs, with eight more 'at-large' making up an extended supporting family.

MH did not begin with the intention of reshaping the community or solving the neighborhood's problems. Instead, we responded to the lack of space for inclusive social opportunities that are not pay-to-play by making one, with a focus on bringing thoughtful, quirky, or difficult programming that had nowhere else to go. In a neighborhood where many have economic limitations, we consciously removed the economic barrier so strangers, friends, acquaintances, and passers-by with little or no disposable income could encounter ideas and practices they may have never seen or considered before. We also wanted to make a space generative of new possibilities.

Without overstating its impact, I want to at least present further theoretical grounding for ways in which I see MH as an 'experimental' cultural center, open resource, and possibility space, standing on the shoulders of Henri Lefebvre (1971 [1991 English translation]). Namely, MH nurtures a rupture in the reproduction of capitalist social relations and space. We are, as much as possible through basic rules, consciously controlling the means of the social production of a space in a rent-free storefront that can nurture capitalism's vast 'outside' – the relations of care, conviviality, and camaraderie that make up our non-economic lives (Gibson-Graham, 1996). I see MH as an attempt to go a step further by creating a space that can enable new relations, spaces, and, in turn, cycles, to emerge. Exploring the spatial manifestations of other such cracks and fissures in the reproduction of capitalism (especially at the concrete scale of everyday life) is perhaps an entire research agenda in itself. Expanding them is a creative practice.

A celebrated 'messiness' of the overlapping yet individual approaches, questions, interests, and assertions permeates both the keyholder project and MH's programming writ large. From the outset, we shared a hope that our approach would create an even bigger mess of overlapping desires, though each keyholder remains free to describe MH and its significance in his or her own way. In the first 'What is Mess Hall?' booklet, which we prepared for our opening night, Brett Bloom described 'a conjoined mess: a multiplicity of significance not reduced for easy consumption.' Sam Gould described MH as

> a conduit. A node. A transmitter. A place for people to beam their thoughts, ideas, and projects out into the world. As well, a point of contact for the people who are searching out these thoughts and ideas ... A shared frequency. A starting point to get the word out, and get a move-on.

Dan S. Wang closed his essay with these assertions:

> In spite of the global interstate system, we (like the anarchists) don't recognize national borders, and work to punch holes in the walls that keep the rest of the world apart from a neighborhood, a city, and a region. Unlike in the capitalist economy, we collapse the distinctions between labor and product (as do the Amish), and between work and play (as did Aristotle). Therefore our partnerships will at times include friends, family, neighbors, professional colleagues, total strangers, and ostensible enemies – in other words all the kinds of people who populate our lives. MESS HALL is not a board room.

These assertions reflect our initial starting point, some of the shared intentions that are constantly shifting as we assess a proposed project or as new keyholders join to coordinate activities, and as our philosophical leanings collide with the practical realities of programming an open resource. Before assessing our accomplishments, I will explain the internal governance structure and expand upon the basic 'rules' that set the parameters of activity.

Governing an open resource: the internal and external MH project

In an open, intense dialogue that began via e-mail in spring 2003, the first eight keyholders, some of us strangers, developed the basic rules that would define the projects to be pursued in this free space. Unlike those who have keys to Gramercy Park in New York City, we were charged with maintaining an open resource. Since the initial keyholder group was first assembled by invitation from the three members of Temporary Services, we have grown in a rather organic way. We extend the group when it becomes clear that someone who has been very active should be invited to do more, or when it seems time to inject some fresh energy and perspectives. In the past, we have reached out to regular attendees at events who show a sustained passion for the place or who are suggested by other keyholders. We are sensitive to maintaining a balance of gender and representation from all races and generations. A new keyholder usually does not know all the other keyholders.

We are an ever-expanding and contracting group, with some keyholders inactive for periods of time that they are unavailable. We are non-hierarchical, sharing responsibilities for opening and closing doors for events, welcoming visitors, answering the frequently asked 'What is this place?' and other questions, making sure the atmosphere of respect and mutual trust is upheld, splitting basic utility costs, and cleaning the bathroom. The governance structure remains an ongoing project, one that runs parallel to, and in many ways mirrors, our approach to programming (Wang, 2004).

That all of us had experience working in collaborative groups, and many with each other in different combinations, informed our desire to *minimize group process* at the start. Keyholders participate as individuals, not as members or representatives of any other group. We all have day jobs, primary artistic practices, and MH, so we want to be efficient with our time while aiming to maximize the resource. We leave room for a multiplicity of voices, approaches, and disagreements, making for what a few of us refer to as a *dissensus model* of organizing. Mutual trust and respect is the baseline for our interaction, for how we treat each other and how we expect visitors to behave. There is no forced unity beyond a shared responsibility to each other and the mission to activate the space for non-monetary forms of exchange. Key to this principle: each of us can bring projects to the group, but unless we explicitly seek assistance, or someone offers, it remains each individual responsibility to execute that project; it does not become something that is realized collectively.

We also have a *flexible participation structure*. Because each keyholder has different commitments and things to contribute, keyholders can participate as much or as little as possible. Whatever energy, time, money, and interests someone can bring to the table are welcome. As noted above, we share the operating costs (at first, between eight people it was approximately $30 each per month) to cover utilities, with people contributing more or less as they are able. Organizers proposing events are responsible for bringing resources to the table in order to execute the project, whether that means mounting an exhibition or having food at an event. Here too, the economy of generosity comes into play with in-kind donations and many projects that collect and repurpose surplus materials.

We consciously have not incorporated as a non-profit. We do not want to be beholden to the governance structure that imposes, nor do we want to pay for it. We also do not want to take energy away from organizing events and programs to write grants. We receive the occasional anonymous donation, and whatever money we make from giving lectures about MH, or from university classes that come to visit, goes to offset operating costs. Those talks are crucial to paying our heating bills in the long Chicago winters.

Growth

At the start, we agreed to organize around food; eating together was enough of a reason for an event. We held lectures and some exhibitions at first, keeping open hours on the weekends for the first six months until it became clear that the real draw was events. We stopped having open hours and instead put time into organizing programs. The power of gathering people together simply to share ideas, food, skills, and miscellaneous items has circulated a surplus of knowledge, cash, social networks, skills, or eclectic interests that fills the schedule and the audiences. The core 'rules' discussed above have survived as new generations of keyholders join. We are, however, presently re-evaluating the organizing structure, as we try to find an efficient way to fill more of the empty spaces in the calendar, particularly in the daytime.

It is clear that the flexible participation principle comes into conflict with the responsibility to maximize the use of the space, and the keyholders' everyday lives. Like others, I would like to see the schedule packed with a diverse mix of activities, but the present structure does not allow for that. With seven keyholders now living in the neighborhood, and a few of them interested in extending the use of the resource to more neighbors on a regular schedule, this prospect is promising.

As part of our fifth anniversary celebration in 2008, we distributed postcards that said 'Your ideas here' around the neighborhood, and asked all who have hosted or attended an event to share thoughts about the past and future of this resource and what it means to them. The responses are feeding our ongoing discussions and upcoming projects.

In short, the model has worked very well, but our development has been somewhat disjointed as the momentum has ebbed and flowed with keyholder commitment. The introduction of so many new keyholders in August 2008 and the exit of others provided a chance to revisit what works, and what does not. As new keyholders have settled in, a palpable enthusiasm for what MH supports and what it can support in the future is generating a lot of ideas and events. Nonetheless, organizational development in an 'organization' that does not really identify as one – and where participation is flexible – poses an interesting challenge, as well as an opportunity. The conversation between keyholders, which now happens on an online forum or via e-mail, is maintained by a handful of active people, with others piping up periodically. We have 13 keyholders but I am left thinking we could have more. I would like to see MH test the outer limits and see how many people and what kind of structure it would take to fully activate the resource.

In the early years we were a unique resource, filling a particular niche and presenting a model that others could follow. Other like-minded spaces have opened since, inspired by elements of MH and its organizing structure, but initiating their own unique experiments. These include InCUBATE (the Institute for Community Understanding Between Art and the Everyday) in a near west-side storefront in Logan Square, and the Monk Parakeet group and Backstory Café in the Experimental Station on the South Side of Chicago in Woodlawn. They provide space for events that MH no longer needs to host. They also provide storefront resources for their local communities in ways MH does not because we do not keep 'business' hours, and are not developing an economy. Yet in its current evolution, MH in Rogers Park is exploring how to further tap the curiosities of its neighbors and stimulate solidarity economies.

Continued dependence on the landlord's generosity is an obvious limitation of MH. Yet Goldberg's long-term commitment to providing the rent-free storefront has allowed a culture – and a cultural economy – of generosity and conviviality to grow and evolve. At the beginning we wondered if or when he would stop donating it. At the time, we agreed that we would want to find another space, but never developed a plan because we never felt a serious threat. Goldberg, for his part, has never applied any pressure. As a result, MH remains focused on the present and immediate future. Goldberg's stated commitment to keeping that storefront a dedicated art space, however, presents a chance to have MH's principles evolve there over the long term. In the meantime, it continues to inspire both visitors and my approach to neighborhood development as I work on ways to make similarly dedicated 'free' space in conjunction with or adjacent to a viable economy that generates livelihoods for those who use it.

MH has a model for programming and organizing that has proven flexible enough to be durable. With a few basic rules – the mutual trust, respect, and latitude I spoke of above deserves additional emphasis – MH is capable of absorbing and being changed by new approaches and ideas. In this way, we model what we are organized to spark. It is also, importantly, a structure that can, and does, outlast any individual participant. Any core or founding group in an organization will decompose over time as people's priorities shift, but MH shows that that too can be accepted as a basic rule. It is perhaps our biggest accomplishment that the spirit of the place, and its purpose, have survived beyond the participation of the founders. This success leaves us in an interesting position to develop a mechanism for institutional memory where the institution is thin, to refine the 'basic rules' and a functional governance structure that is anything but bureaucratic.

Towards the distributional imperative

My purpose in detailing the MH case is, on the one hand, to highlight how this space outside of monetary exchanges – a decommodified storefront – functions and sustains itself with an economy of generosity and conviviality. In a space where newcomers are known to ask 'What do you sell here?' MH offers a reorientation towards social activities – from gathering to eat, to collaborating on projects. We do not sell anything; we give things (and non-things) away. MH shows that generosity can be generative and dynamic. We learn, we share, we listen, we argue, we create, we think, we experiment. These unquantifiably valuable activities lack sufficiently open and accessible space to flourish outside our homes. They may not pay the rent, but they nonetheless constitute a vital part of human communities, and can be thought of as a public good. In this final section, I will introduce the *distributional imperative* of the so-called 'new,' knowledge, or creative economy and draw out some lessons from MH to approach it.

The *distributional imperative* is nothing less than the imperative to develop an accessible infrastructure for full participation in the fruits of technological, social, and cultural advances of our age and in the chance to make new ones. Approaching this challenge might mean making *access points*, 'free' space where diverse economic possibilities, the seeds of deeper democracy, and self-education can thrive. How can we reinsert unpredictability in urban places where it has been designed out of existence? A place like Mess Hall offers one approach, a physical site. Just as the open-source movement argues for expanding the knowledge commons – that making more knowledge available to be recombined, reconfigured, and reconstituted as something else entirely can be to the collective benefit of all – I would make a case for a socio-spatial commons, possibility spaces, places where people can meet, linger, and 'be together.' These are not so-called 'public spaces' surveyed by anonymous cameras. They are places that are cared for, where people are cared for; such relations make surveillance cameras redundant. They are social worlds, the sites of simultaneous and multi-dimensional engagement, the eventual, myriad offline reasons we might connect online.

The distributional imperative puts a finer point on the potential sites of invention in cities, and who has access to them. For, as Allan Scott writes at the close of his 2006 essay:

In the last analysis, any push to achieve urban creativity in the absence of a wider concern for conviviality and camaraderie (which need to be distinguished from the mechanical conception of diversity) in the urban community as a whole is doomed to remain radically unfinished.

(Scott, 2006: 15)

Taking Scott's point seriously means creating environments where people can develop creative capacities where they live. It suggests basic questions: Where are learning environments and skills for creative vocations or creative critical thinking accessed? Might there be one-stop creative vocational centers? What would happen there? Where would they be located, and by whom (and how) would they be governed and financed? As the MH case suggests, the space to innovate begins in the organization itself, and the content and connections it supports.

Celebrating, describing, and defending how to measure the value of creative work after economic restructuring has been a big theme in scholarship and policy since Richard Florida's *Rise of the Creative Class* in 2002. The preoccupation with assessing the economic productivity of creative work may well have obscured the need for deeper work on non-economic and non-accumulation-oriented creativities. Further research – that overcomes the methodological shortsightedness of measuring creativity in terms of exchange outputs – is needed to understand what role possibility spaces of various sorts play in stimulating productive surpluses and creating new forms of work.

With this in mind, I suggest Mess Hall points to a rich arena for further research on and experimentation with the generative power of neighborhood possibility spaces, and of economies of generosity and conviviality. MH suggests but one model for considering how a 'thin' institution with consistent access to free space can enable 'thick' encounters among strangers. Creating spaces – close to home, but not home – where strangers and neighbors can share ideas, learn, invent, and collaborate is one way to approach the distributional imperative of the creative economy, enable people to be together in their differences, and spark untold initiatives. Or, in the words of one keyholder during an event in February 2009, 'there are a whole bunch of people I don't know working on a project together, and I don't know who started the project. Must be a Mess Hall moment.' In a context where the provision of substantively open and public space is increasingly under threat, but disused spaces are everywhere, the challenge rests in enabling more moments like these.

References

ACE (Arts Council of England) (2006) *Arts Centres Research*, London: Arts Council of England, September.

Active 18 (2008) Available at http://active18.org.

Akimbo (2008) 'Mammalian Diving Reflex Presents', Akimbo Community Events Posting, www.akimbo.ca/events/?id=11542&day=16&month=6&year=2008.

Allen, J. and Cars, G. (2001) 'Multiculturalism and governing neighbourhoods', *Urban Studies*, 38(12): 195–209.

Alvarez, M. (2005) *There's Nothing Informal About It: Participatory Arts Within the Cultural Ecology of Silicon Valley*, San Jose, CA: Cultural Initiatives Silicon Valley.

Amélie (2001) Dir: Jeunet, J-P., Momentum Pictures.

American Gnome Liberationist Site (2008) Available at www.freethegnomes.com/index.html.

Americans for the Arts (2002) *Arts and Economic Prosperity: The Economic Impact of Nonprofit Arts Organizations and Their Audiences*, Washington, DC: Americans for the Arts.

Anthes, B. (2006) *Native Moderns: American Indian Painting, 1940–1960*, Durham: Duke University Press.

Appadurai, A. (ed.) (1986) *The Social Life of Things: Commodities in Cultural Perspective*, Cambridge: Cambridge University Press.

Arendt, H. (1958) *The Human Condition*, Chicago: University of Chicago Press.

Artscape (2008) 'Queen West Art Crawl'. Available at www.torontoartscape.on.ca/qwac/.

Asheim, B., Coenen, L. and Vang, J. (2007) 'Face-to-face, buzz, and knowledge bases: sociospatial implications for learning, innovation, and innovation policy', *Environment and Planning C: Government and Policy*, 25(5): 655–70.

Atkinson, D. and Laurier, E. (1998) 'A sanitised city? Social exclusion at Bristol's 1996 International Festival of the Sea', *Geoforum*, 29(2): 199–206.

Atkinson, N. (2007) 'Le chic bilingue', *National Post*, 24 March: WP.3.

Attfield, J. (2000) *Wild Things: The Material Culture of Everyday Life*, Oxford: Berg.

Attfield, J. (2006) 'Redefining kitsch: the politics of design', *Home Cultures*, 3: 210–12.

Australian Bureau of Statistics (ABS) (2001) *Census of Population and Housing*, Canberra: ABS.

Bachelard, G. (2000) *The Dialectics of Duration*, Manchester: Clinamen Press.

Bain, A. (2003) 'Constructing contemporary artistic identities in Toronto neighbourhoods', *The Canadian Geographer*, 47(3): 303–17.

Bain, A. (2006) 'Resisting the creation of forgotten places: artistic production in Toronto neighbourhoods', *The Canadian Geographer*, 50(4): 417–31.

Bain, A. (forthcoming) 'Re-imaging, re-elevating, re-placing the urban', in T. Bunting, P. Filion, and R. Walker (eds), *Canadian Cities in Transition: New Directions in the 21st Century*, Oxford: Oxford University Press.

Balzer, D. (2007) 'Fear and loathing and public art: Chris Reynolds and Elaine Gaito encourage play in Parkdale', *Eye Weekly.* Available at: www.eyeweekly.com/features/article/1363.

Banks, M. (2009) 'Fit and working again? The instrumental leisure of the "creative class"', *Environment and Planning A*, 41(3): 668–81.

Banks, M., Lovatt, A., O'Connor, J. and Raffo, C. (2000) 'Risk and trust in the cultural industries', *Geoforum*, 31(4): 453–64.

Barker, J. and Weller, S. (2002) 'Geography of methodological issues in research with children', *Qualitative Research*, 3(2): 207–27.

Barnes, C. and Land, D. (2007) 'Geographies of generosity: beyond the "moral turn"', *Geoforum*, 38: 1065–75.

Barnes, K., Waitt, G., Gill, N. and Gibson, C. (2006) 'Community and nostalgia in urban revitalisation: a critique of urban village and creative class strategies as remedies for social "problems"', *Australian Geographer*, 37(3): 335–54.

Barney, D. (2004) *The Network Society*, Cambridge: Polity Press.

Barrett, H. and Phillips, J. (1993) *Suburban Style: The British Home, 1840–1960*, London: Little, Brown.

Barry, A. (1999) 'Invention and inertia', *Cambridge Anthropology*, 21: 62–70.

Batchen, G. (2001) *Each Wild Idea: Writing, Photography, History*, Cambridge, MA: MIT Press.

Batchen, G. (2002) 'On the history of photography: a talk with Geoffrey Batchen', *Folio.* Available at www.gc.cuny.edu/faculty/folio/fall2002/Batchen.htm.

Bathelt, H. and Malmberg, A. (2004) 'Clusters and knowledge: local buzz, global pipelines and the process of knowledge creation', *Progress in Human Geography*, 28(1): 31–56.

Baudrillard, J. (1996) *The System of Objects*, London: Verso.

Bauman, Z. (2000) *Liquid Modernity*, Cambridge: Polity Press.

BBC News (2004) 'Garden gnomes culled at waste tip', April 27. Available at http://news.bbc.co.uk/2/hi/uk_news/england/west_midlands/3665067.stm.

BBC News (2007) 'Woman faces jail over gnome theft'. Available at http://news.bbc.co.uk/1/hi/scotland/tayside_and_central/6500309.stm.

Becker, H. (1982) *Art Worlds*, Berkeley: University of California Press.

Bell, D. and Hollows, J. (eds) (2005) *Ordinary Lifestyles: Popular Media, Consumption and Taste*, Maidenhead: Open University Press.

Bell, D. and Jayne, M. (2003) 'Design-led urban regeneration: a critical perspective', *Local Economy*, 18(2): 121–34.

Bell, D. and Jayne, M. (eds) (2004) *City of Quarters: Urban Villages in the Contemporary City*, Aldershot: Ashgate.

Bell, D. and Jayne, M. (2006) 'Conceptualising small cities', in D. Bell and M. Jayne (eds), *Small Cities: Urban Experience Beyond the Metropolis*, London: Routledge.

Bell, D. and Jayne, M. (forthcoming) 'The creative countryside: policy and practice in the rural cultural economy', *Geoforum*.

Belk, R.W. (1987) 'A child's Christmas in America: Santa Claus as deity, consumption as religion', *Journal of American Culture*, 10: 87–100.

Benjamin, W. (2002) *The Arcades Project*, ed. Rolf Tiedemann, trans. Howard Eiland and Kevin McLaughlin, New York: Belknap Press.

Benjamin, W. (2003) 'On the concept of history', in *Selected Writings: 1938–1940, Volume 4*, Cambridge, MA: Harvard University Press.

Benkler, Y. (2006) *The Wealth of Networks: How Social Production Transforms Markets and Freedom*. Available at www.benkler.org.

Bergson, H. (1911) *Creative Evolution*, trans. A. Mitchell, London: Macmillan.

Berlo, J. and Phillips, R. (1998) *Native North American Art*, Oxford: Oxford University Press.

Berman, M. (1983) *All That Is Solid Melts Into Air: The Experience of Modernity*, London: Verso.

Bertolini, L. and Salet, W. (2003) 'Planning concepts for cities in transition: regionalization of urbanity in the Amsterdam structure plan', *Planning Theory and Practice*, 4(2): 131–46.

Bhabha, H. (1994) *The Location of Culture*, London: Routledge.

Bhabha, H. (1990) 'The third space: interview with Homi K. Bhabha', in J. Rutherford (ed.) *Identity: Community, Culture, Difference*, London: Lawrence & Wishart, 207–21.

Bhatti, M. and Church, A. (2001) 'Cultivating natures: homes and gardens in late modernity', *Sociology*, 35: 365–83.

Bill, A. (2008) *Creative Girls: Fashion Design Education and Governmentality*, PhD thesis, Auckland: University of Auckland.

Binkley, S. (2000) 'Kitsch as a repetitive system: a problem for the theory of taste hierarchy', *Journal of Material Culture*, 5: 131–52.

Bird, J. (1993) 'Dystopia on the Thames', in J. Bird, B. Curtis, T. Putnam, G. Robertson and L. Tickner (eds), *Mapping the Futures: Local Cultures, Global Change*, London: Routledge.

Blomley, N. (2007) 'Critical geography: anger and hope', *Progress in Human Geography*, 31(1): 53–65.

Bloom, B. (2006) 'Radical space for art in a time of forced privatization and market dominance', paper presented at Wir Sind Woanders Conference. Hamburg, Germany, November.

Blue Velvet (1986) Dir. D. Lynch, De Laurentis Entertainment.

Boden, M. (1990) *The Creative Mind: Myths and Mechanisms*, London: Weidenfeld and Nicolson.

Bohemian Embassy (2008) Available at http://bohemian embassy.ca.

Bonnett, A. (1996) 'The transgressive geographies of everyday life: socialist pathways within everyday urban creativity', *Transgressions*, 2–3: 20–37.

Borthwick, S. and Moy, R. (2004) *Popular Music Genres: An Introduction*, London and New York: Routledge.

Bourdieu, P. (1976) *Distinction: A Social Critique of the Judgement of Taste*, London: Routledge and Kegan Paul.

Bourdieu, P. (1993) *The Field of Cultural Production: Essays on Art and Literature*, Cambridge: Polity Press.

Bourdieu, P. and Darbel, A. (1991) *The Love of Art*, Cambridge: Polity Press.

Bouton, L. (2007) 'Power and perception: Pablita Velarde', *Indian Market*, August: 184.

Bowie, D. (1967) 'The laughing gnome', 7 inch single, DM 123.

Boyle, M. (1997) 'Civic boosterism in the politics of local economic development: "institutional positions" and "strategic orientations" in the consumption of hallmark events', *Environment and Planning A*, 29(11): 1975–97.

Brennan-Horley, C. and Gibson, C. (forthcoming) 'Where is creativity in the city? Integrating qualitative and GIS methods', *Environment and Planning A*, accepted for publication Dec 23, 2008.

Brennan-Horley, C., Connell, J. and Gibson, C. (2007) 'The Parkes Elvis Revival Festival: economic development and contested place identities in rural Australia', *Geographical Research*, 45(1): 71–84.

Brenner, N. (2004) *New State Spaces: Urban Governance and the Rescaling of Statehood*, Oxford and New York: Oxford University Press.

Brenner, N. and Theodore, N. (2002) 'Cities and the geographies of "actually existing neoliberalisms"', *Antipode*, 34(3): 349–79.

Brenner, N. and Theodore, N. (eds) (2002) *Spaces of Neoliberalism: Urban Restructuring in North America and Western Europe*, Oxford: Blackwell.

Bröckling, U. (2007) *Das unternehmerische Selbst: Soziologie einer Subjektivierungsform*. Frankfurt am Main: Suhrkamp.

Brody, J. (1976) 'The creative consumer: survival, revival and invention in Southwest Indian Arts', in N. Grayburn (ed.), *Ethnic and Tourist Arts: Cultural Expressions from the Fourth World*, Berkeley: University of California Press.

Bruce Museum (2007) 'Press release: contemporary and cutting edge: pleasures of collecting, Part III September 29, 2007 to January 6, 2008'. Available at www.brucemuseum.org/aboutus/press/ContemporaryandCuttingEdgepublicprograms.doc.

Buchan, U. (2000) 'Frank's fame' *The Spectator*, January 22.

Buchli, V. (2002) 'Introduction', in V. Buchli (ed.), *The Material Culture Reader*, Oxford: Berg.

Buffy the Vampire Slayer (1999) 'The Zeppo', Season 3, episode 13. Dir. James Whitmore, Jr.

Bunting, T. and Mitchell, C. (2001) 'Artists in rural locales: market access, landscape appeal, and economic exigency', *Canadian Geographer*, 45(2): 268–84.

Bunting, T., Walks, A. and Filion, P. (2004) 'The uneven geography of housing affordability stress in Canadian metropolitan areas', *Housing Studies*, 19(3): 361–93.

Bunting, C., Chain, T., Goldthorpe, J., Keany, E. and Oskala, A. (2008) *From Indifference to Enthusiasm: Patterns of Arts Attendance in England*, London: Arts Council of England.

Burgess, E. (2003) 'The growth of a city: an introduction to a research project', in R.T. LeGates and F. Stout (eds), *The City Reader*, London: Routledge, 153–61.

Burgess, J. (2006) 'Hearing ordinary voices: cultural studies, vernacular creativity and digital storytelling', *Continuum: Journal of Media & Cultural Studies*, 20(2): 201–14.

Burgess, J. (2007) *Vernacular Creativity and New Media*, PhD thesis, Brisbane: Queensland University of Technology.

Burningham, K. and Thrush, D. (2001) *'Rainforests Are a Long Way From Here': The Environmental Concerns of Disadvantaged Groups*, York: Joseph Rowntree Foundation.

Burrows, R. and Rhodes, D. (1998) *Unpopular Places? Area Disadvantage and the Geography of Misery*, Bristol: Policy Press.

Butler, J. (1989) 'Performative acts and gender constitution: an essay in phenomenology and feminist theory', in S. Case (ed.), *Performing Feminisms: Feminist Critical Theory and Theatre*, Baltimore: Johns Hopkins University Press.

Butterfield, S. (2006) 'Eyes of the World', *FlickrBlog*. Available at http://blog.flickr.com/flickrblog/2006/03/eyes_of_the_wor.html.

CBC (Canadian Broadcasting Corporation) (2007) 'Budget crisis triggers widespread cuts to Toronto city services', CBC News On-Line. Available at www.cbc.ca/canada/toronto/story/2007/08/10/toronto-budget.html.

Carlsson, C. (2008) *Nowtopia: How Pirate Programmers, Outlaw Bicyclists, and Vacant-lot Gardeners are Inventing the Future Today*, Edinburgh/Oakland: AK Press.

Castaner, X. and Campos, L. (2002) 'The determinants of artistic innovation: bringing in the role of organizations', *Journal of Cultural Economics*, 26: 29–52.

Castells, M. (1996) *The Rise of the Network Society*, Blackwell: Oxford.

Caves, R. (2000) *Creative Industries: Contracts between Art and Commerce*, Cambridge, MA: Harvard University Press.

CEM (Culture East Midlands) (2007) *Creating Cultural Opportunites in Sustainable Communities*, Nottingham: CEM. Available at: www.culture-em.org.uk/documents/uploads/Creating%20Cultural%20Opps%20(4)1.pdf.

Cheal, D. (1988) *The Gift Economy*, London: Routledge.

Cheney, P. (2007) 'The humble legend', *The Globe & Mail*, 26 December, A1, A18.

Christophers, B. (2008) 'The BBC, the creative class, and neoliberal urbanism in the north of England', *Environment and Planning A*, 40(10): 2313–29.

Church, C. and Elster, J. (2002) *Thinking Locally, Acting Nationally: Lessons for National Policy from Work on Local Sustainability*, York: Joseph Rowntree Foundation.

City of Montreal (2003) 'Framework, guiding principles and statement for a cultural policy', *Summary of the Report of the Advisory Group*, Montréal: City of Montréal.

City of Montreal (2005a) *Reussir@Montreal. Stratégie de développement économique 2005–2010 de la Ville de Montréal*, Montréal: City of Montréal.

City of Montreal (2005b) *Nouvelles villes de design/New Design Cities*, Montréal: Infopresse.

City of Montreal (2005c) *Imagining – Building Montréal 2025 – a World of Creativity and Opportunities*, Montréal: City of Montréal.

City of Montreal (2006) *Montréal Ville Unesco de Design/Unesco City of Design*, Montréal: City of Montréal.

City of Toronto (2005) *Culture Plan for the Creative City*, Available at www.toronto.ca/culture/cultureplan.htm.

Clark, M. (2006) 'Grimes düsterer Bruder: Dubstep', *De:Bug*, 28.07. Available at www.de-bug.de/mag/index.php?ID=4281.

Coates, T. (2006) 'What do we do with "social media"?' *Plastic Bag*. Available at: www.plasticbag.org/archives/2006/03/what_do_we_do_with_social_media.shtml.

Cohen, E. (1999) 'Cultural fusion', in *Values and Heritage Conservation*, Los Angeles: Getty Conservation Institute.

Cohen, S. (2007) *Decline, Renewal and the City in Popular Music Culture: Beyond the Beatles*, Aldershot: Ashgate.

Coles, R. (1997) *Rethinking Generosity: Critical Theory and the Politics of Caritas*. Ithaca, NY: Cornell University Press.

Collings, M. (1999) *This is Modern Art*, London: Weidenfeld & Nicolson.

Cotter, H. (2003) 'Doing their thing, making art together', *New York Times*, 19 January.

Cowen, D. (2006) 'Hipster urbanism', *Relay: A Socialist Project Review*, issue 13, September/October.

Cox, G. (2005) *Cox Review of Creativity in Business: Building on the UK's Strengths*, London: HM Treasury.

Crang, M. (2003) 'On display: the poetics, politics and interpretation of exhibitions', in A. Blunt, P. Gruffudd, J. May, M. Ogborn and D. Pinder (eds), *Cultural Geography in Practice*, London: Hodder Arnold.

Creative London (2003) *Creative London – Vision and Plan*, London: Creative London.

Cresswell, T. (1996) *In Place, Out of Place: Geography, Ideology and Transgression*, Minneapolis: Minnesotoa University Press.

Crewe, L. and Beaverstock, J. (1998) 'Fashioning the city: cultures of consumption in contemporary urban spaces', *Geoforum*, 29(3): 287–308.

Crouch, D. (1989) 'Patterns of co-operation in the cultures of outdoor leisure – the case of the allotment', *Leisure Studies*, 8(2): 189–99.

Crouch, D. (2003a) 'Spacing, performance and becoming: the tangle of the mundane', *Environment and Planning A*, 35: 1945–60.

Crouch, D. (2003b) 'Performance and the constitutions of nature: a consideration of the performance of lay geographies', in B. Szerszynski, W. Heim and C. Waterton (eds), *Natures Performed*, Oxford: Blackwell.

Crouch, D. (2003c) *The Art of Allotments*, Nottingham: Five Leaves.

Crouch, D. (2009) 'Gardening and gardens', in R. Kitchen and N. Thrift (eds) *International Encyclopaedia of Human Geography*, Norwich: Elsevier.

Crouch, D. and Toogood, M. (1999) 'Everyday abstraction: geographical knowledge in the art of Peter Lanyon', *Ecumene* 6(1): 72–89.

Crouch, D. and Ward, C. (1988) *The Allotment: Its Landscape and Culture*, London: Faber and Faber.

Crozier, M. (2003) 'Simultanagnosia, sense of place and the garden idea', *Thesis Eleven*, 74: 76–88.

Csikszentmihalyi, M. (1997) *Creativity: Flow and the Psychology of Discovery and Invention*, New York: HarperCollins.

Currid, E. (2007a) *The Warhol Economy: How Fashion, Art and Music Drive New York City*, Princeton: Princeton University Press.

Currid, E. (2007b) 'How art and culture happens in urban economies: implications for the economic development of culture', *Journal of the American Planning Association*, 73(4): 454–67.

Davidoff, P. (1965) 'Advocacy and pluralism in planning', *Journal of the American Institute of Planning*, 31(4): 331–8.

DCMS (Department for Culture, Media and Sport) (1998) *Creative Industries Mapping Document*, London: Department for Culture, Media and Sport.

DCMS (2007) *Taking Part: The National Survey of Culture, Leisure and Sport*, Provisional results from the first six months of the 2005/6 Survey, London: Department for Culture, Media and Sport.

DCMS (2008) *Our Creative Talent: The Voluntary and Amateur Arts in England*, London: Department for Culture, Media and Sport.

de Certeau, M. (1984) *The Practice of Everyday Life*, Berkeley: University of California Press.

Degen, M. (2008) *Sensing Cities: Regenerating Public Life in Barcelona and Manchester*, London: Routledge.

Deitch, L. (1989) 'The impact of tourism on the arts and crafts of the Indians of the Southwestern United States', in V. Smith (ed.), *Hosts and Guests: The Anthropology of Tourism, Second Edition*, Philadelphia: University of Pennsylvania Press.

DeLanda, M. (2004) 'Space: extensive and intensive, actual and virtual', in I. Buchannan and G. Lambert (eds), *Deleuze and Space*, London: Routledge.

Deleuze, G. and Guattari, F. (2004) *A Thousand Plateaus*, London: Continuum.

Derrida, J. (1992) *Given Time*, trans. Peggy Kamuf, Chicago: University of Chicago Press.

Design Museum, Gent (2007) 'Kitsch, camp or design?'. Available at http://design.museum. gent.be/ENG/exhibitions-archive/kitsch_e.php.

DETR (Department of the Environment, Transport and the Regions) (2000) *Our Towns and Cities: The Future. Delivering an Urban Renaissance*, government White Paper, London: HMSO.

Dewsbury, J.D. and Thrift, N. (2004) '"Genesis eternal": after Paul Klee', in I. Buchanan and G. Lambert (eds), *Deleuze and Space*, Edinburgh: Edinburgh University Press.

DIAC (Design Industry Advisory Committee) (2004) *What Can 40,000 Designers do for Ontario? Executive Report*, Toronto: DIAC.

Dilworth, L. (1996) *Imagining Indians in the Southwest: Persistent Visions of a Primitive Past*, Washington, DC: Smithsonian Institution Press.

Diprose, R. (2002) *Corporeal Generosity: On Giving with Nietzsche, Merleau-Ponty, and Levinas*, Albany: State University of New York Press.

Donald, B. and Blay-Palmer, A. (2006) 'The urban creative-food economy: producing food for the urban elite or social inclusion opportunity?', *Environment and Planning A*, 38(10): 1901–20.

Dorfles, G. (1968) *Kitsch; the World of Bad Taste*, New York: Universe Books.

Drake Hotel (2008) Available at www.thedrakehotel.ca/.

Dreher, C. (2002) 'Be creative – or die', *Salon*, 6 June: 1.

Dunn, P. and Leeson, L. (1993) 'The art of change in Docklands', in J, Bird, B. Curtis, T. Putnam, G. Robertson and L. Tickner (eds), *Mapping the Futures: Local Cultures, Global Change*, London: Routledge: 138–50.

Eagleton, T. (2000) *The Idea of Culture*, Oxford: Blackwell.

EDAW (2007) *Report of City Centre Masterplan 2007. Review and Roll Forward*, Comments from Public Exhibition: 29 November–2 December 2006, Sheffield: Sheffield City Council.

Edensor, T. (2005) *Industrial Ruins: Space, Aesthetics and Materiality*, Oxford: Berg.

Edensor, T. and Millington, S. (2009) 'Illuminations, class identities and the contested landscapes of Christmas', *Sociology*, 43(1): 103–21.

Edwards, P. (1999) 'Culture is ordinary – Raymond Williams and cultural materialism', *Red Pepper*, August. Available at www.users.zetnet.co.uk/amroth/scritti/williams.htm.

Erasure (2003) Video for 'Make Me Smile'. Available at www.youtube.com/watch?gl=GB&v=uGcm4k2R2Ww.

Evans, G. (1998) *Study into the Employment Effects of Arts Lottery Spending in England*, Research Report No. 14, London: Arts Council of England.

Evans, G. (2001) *Cultural Planning: An Urban Renaissance?* London: Routledge.

Evans, G. (2003) 'Hard branding the culture city – from Prado to Prada', *International Journal of Urban and Regional Research*, 27(2): 417–40.

Evans, G. (2004) 'Country profiles: United Kingdom', in C. Bodo, C. Gordon and D. Ilczuk (eds), *Gambling on Culture. State Lotteries as a Source of Funding for Culture – the Arts and Heritage*, Amsterdam: CIRCLE/Associazione per L'Economica della Cultura/Boekmanstudies.

Evans, G. (2005) 'Measure for measure: evaluating the evidence of culture's contribution to regeneration', *Urban Studies*, 42(5/6): 959–84.

Evans, G. (2007) 'Tourism, creativity and the city', in G. Richards and J. Wilson (eds), *Tourism Creativity and Development*, London: Routledge.

Evans, G. (2009) 'Creative cities, creative spaces and urban policy', *Urban Studies*, 46(5–6): 1003–40.

Evans, G. and Foord, J. (2000) 'European funding of culture: promoting common culture or regional growth?', *Cultural Trends*, 36: 53–87.

Evans, G. and Foord, J. (2004) 'Rich mix cities: from multicultural experience to cosmopolitan engagement', *Ethnologia Europaea: Journal of European Ethnology*, 34(2): 71–84.

Evans, G. and Foord, J. (2006) 'Small cities for a small planet: sustaining the urban renaissance', in M. Jayne and D. Bell (eds) *Small Cities: Urban Experience Beyond the Metropolis*, London: Routledge.

Evans, G. and Foord, J. (2008) 'Cultural mapping and sustainable communities: planning for the arts revisited', *Cultural Trends*, 17(2): 65–96.

Evans, K. and Lowery, D. (2006) 'The impact of visual and expressive art on public policy and public voice', paper presented at Drawing the Lines: International Perspectives on Urban Renewal Through the Arts, Indiana University Northwest, Gary, Indiana, 2–4 November.

Eversole, R (2005) 'Challenging the creative class: innovation, "creative regions" and community development', *Australasian Journal of Regional Studies*, 11(3): 351–60.

Featherstone, M. (1991) *Consumer Culture and Postmodernism*, London: Sage.

Featherstone, M. (1992) 'Heroic life and everyday life', in M. Featherstone (ed.), *Cultural Theory and Cultural Change*, London: Sage.

Ferris, J., Norman, C. and Sempik, J. (2001) 'People, land and sustainability: community gardens and the social dimension of sustainable development', *Social Policy and Administration*, 35(5): 559–68.

Fine, G. (2003) 'Crafting authenticity: the validation of identity in self-taught art', *Theory and Society*, 32: 153–80.

Fine, G. (2004) *Everyday Genius: Self-Taught Art and the Culture of Authenticity*, Chicago: Chicago University Press.

Finnegan, R. (1989) *The Hidden Musicians: Music-Making in an English Town*, Cambridge: Cambridge University Press.

Florida, R. (2002) *The Rise of the Creative Class: And How it's Transforming Work, Leisure, Community and Everyday Life*, New York: Perseus Books Group.

Florida, R. (2005) *Cities and the Creative Class*, London: Routledge.

Forman, M. (2000) '"Represent": race, space and place in rap music', *Popular Music*, 19: 65–90.

Foucault, M. (1986) 'Of other spaces', *Diacritics*, 16(1): 22–7.

Foucault, M. (1991) 'Governmentality', in G. Burchell, C. Gordon and P. Miller (eds), *The Foucault Effect. Studies in Governmentality*, Chicago: University of Chicago Press.

Foxnews.com (2005) 'Garden gnome meets Paris Hilton', 18 April. Available at www.foxnews.com/story/0,2933,153715,00.html.

Franklin, A. (2001) *Nature and Social Theory*, London: Sage.

Fraser, N. (1997) *Justice Interruptus: Critical Reflections on the Postsocialist Condition*, New York: Routledge.

Fraser, N. (2003) *Redistribution or Recognition? A Political-Philosophical Exchange*, London: Verso.

Frey, B. (1999) 'State support and creativity in the arts: some new considerations', *Journal of Cultural Economics*, 23(1): 71–85.

Friedmann, J. (1987) *Planning in the Public Domain: From Knowledge to Action*, Princeton: Princeton University Press.

Friedmann, J. (2007) 'The invisible web: place and place-making in cities'. Available at www.scarp.ubc.ca/faculty%20profiles/friedmannpubs.htm.

Friends of the Earth (2001) *Pollution and Poverty – Breaking the Link*, London: Friends of the Earth.

Garlake, M. (1995) 'Peter Lanyon's letters to Naum Gabo', *Burlington Magazine*, April.

Garnham, N. (2001) 'Afterword: the cultural commodity and cultural policy', in S. Selwood (ed.), *The UK Cultural Sector*, London: Policy Studies Institute.

Germain, A. and Rose, D. (2000) *Montréal: The Quest for a Metropolis*, New York: John Wiley and Sons.

Gertler, M. (1995) '"Being there": proximity, organization, and culture in the development and adoption of advanced manufacturing technologies', *Economic Geography*, 71(1): 1–26.

Gertler, M. (2003) 'Tacit knowledge and the economic geography of context, or the undefinable tacitness of being (there)', *Journal of Economic Geography*, 3(1): 75–99.

Gertler, M., Florida, R., Gates, G. and Vinodrai, T. (2002) *Competing on Creativity: Placing Ontario's Cities in a North American Context*, Toronto: Ontario Ministry of Enterprise, Opportunity, and Innovation.

Gibson, C. (2002) 'Rural transformation and cultural industries: popular music on the New South Wales Far North Coast', *Australian Geographical Studies*, 40(3): 336–56.

Gibson, C. (2007) 'Music festivals: transformations in non-metropolitan places, and in creative work', *Media International Australia incorporating Culture and Policy*, 123 (May): 65–81.

Gibson, C. (2009) 'Creative arts, people and places: which policy directions?', in Andersen, L. and Oakley, K. (eds), *How Are We Going? Directions for the Arts in the Creative Age*, Cambridge: Cambridge Scholars Press.

Gibson, C. and Connell, J. (2009) *Music Festivals and Regional Development in Australia*, Aldershot: Ashgate.

Gibson, C. and Davidson, D. (2004) 'Tamworth, Australia's "country music capital": place marketing, rural narratives and resident reactions', *Journal of Rural Studies*, 20(4): 387–404.

Gibson, C. and Klocker, N. (2004) 'Academic publishing as "creative" industry, and recent discourses of "creative economies": some critical reflections', *Area*, 36(4): 423–34.

Gibson, C. and Klocker, N. (2005) 'The "cultural turn" in Australian regional economic development discourse: neoliberalising creativity?', *Geographical Research*, 43(1): 93–102.

Gibson, C. and Kong, L. (2005) 'Cultural economy: a critical review', *Progress in Human Geography*, 29(5): 541–61.

Gibson, C., Waitt, G., Walmsley, J. and Connell, J. (forthcoming) 'Cultural festivals and economic development in regional Australia', *Journal of Planning Education and Research*, conditionally accepted 16/10/08.

Gibson-Graham, J. (1996) *The End of Capitalism (As We Knew It)*, Minneapolis: University of Minnesota Press.

Gibson-Graham, J. (2006) *A Post-Capitalist Politics*, Minneapolis: University of Minnesota Press.

Gibson-Graham, J. (2008) 'Diverse economies: performative practices for "other worlds"' *Progress in Human Geography*, 32(5): 613–32.

Giddens, A. (1994) 'Living in a post-traditional society', in U. Beck, A. Giddens and S. Lash (eds), *Reflexive Modernization*, Cambridge: Polity Press.

Gilbert, J. (2008) *Anticapitalism and Culture: Radical Theory as Popular Politics*, Oxford: Berg.

Gilbert, D. and Preston, R. (2003) 'Stop being so English: suburbia and national identity', in D. Gilbert, D. Matless and B. Short (eds), *Geographies of British Modernity: Space and Society in the Twentieth Century*, Oxford: Blackwell.

Gilbert, J. and Pearson, E. (1999) *Discographies: Dance Music, Culture and the Politics of Sound*, London: Routledge.

Gill, R. (2002) 'Cool, creative and egalitarian? Exploring gender in project-based new media work in Europe', *Information, Communication and Society*, 5(1): 70–89.

GLA (Greater London Authority) (2002) *Creativity: London's Core Business*, London: Greater London Authority.

GLA (2003) *Play it Right: Asian Creative Industries in London*, London: Greater London Authority.

GLA (2004a) *London Cultural Capital. Realising the Potential of a World-Class City*, London: Greater London Authority.

GLA (2004b) *London's Creative Sector: 2004 Update*, London: Greater London Authority.

Gladwell, M. (2000) *The Tipping Point: How Little Things Can Make a Big Difference*, Boston: Little, Brown.

GLC (Greater London Council) (1985) *State of the Arts or the Art of the State: Strategies for the Cultural Industries*, London: Greater London Council.

Glover, T., Shinew, K. and Parry, D. (2005) 'Association, sociability, and civic culture: the democratic effect of community gardening', *Leisure Sciences*, 27(1): 75–92.

Gnome Away from Home. Available at www.amazon.com/Gnome-Away-Andrews-McMeel-Publishing/dp/0740757237.

Gnome Liberation Front (Front de Libération des Nains des Jardins Francais). Available at www.flnjfrance.com/.

Golby, J.M. and Purdue, A.W. (1986) *The Making of Modern Christmas*, London: BT Batsford.

Goldbard, A. (2006) *New Creative Community: The Art of Cultural Development*, Oakland, CA: New Village Press.

Gómez-Peña, G. (2005) *Ethno-Techno, Writings in Performance, Activism and Pedagogy*, New York: Routledge.

Gorman-Murray, A., Waitt, G. and Gibson, C. (2008) 'A queer country? A case study of the politics of gay/lesbian belonging in an Australian country town', *Australian Geographer*, 39(2): 171–91.

Gough, J., Eisenschitz, A. with McCulloch, A. (2006) *Spaces of Social Exclusion*, London: Routledge.

Grabher, G. (2002) 'Cool projects, boring institutions: temporary collaboration in social context', *Regional Studies*, 36, 205–14.

Greenwood, W. (1933) *Love on the Dole*, London: Penguin.

Gregson, N., Metcalfe, A. and Crewe, L. (2007) 'Moving things along: the conduits and practices of divestment in consumption', *Transactions of the Institute of British Geographers*, 32(2): 187–200.

Gritton, J. (2000) *The Institute of American Indian Arts: Modernism and U.S. Indian Policy*, Albuquerque: University of New Mexico Press.

Gross, L. (1995) 'Art and artists on the margins', in L. Gross (ed.), *On the Margins of Art Worlds*. Boulder, CO: Westview Press.

Grosz, E. (1999) 'Thinking the new: of futures yet unthought', in E. Grosz (ed.), *Becomings: Explorations in Time, Memory and Futures*, Ithaca, NY: Cornell University Press.

Groth, P (1997) in P. Groth, and T.W. Bressi (eds), *Understanding Ordinary Cultures*, New Haven, CT: Yale University Press, 1–24.

Groundwork Trust (2001) *From Community Gardening to Westminster*, Birmingham: Groundwork Trust.

Habermas, J. (1996) *Between Facts and Norms: Contributions to a Discourse Theory of Law and Democracy*, Cambridge: Polity Press.

Hall, C. (1997) 'Mega-events and their legacies', in P. Murphy (ed.), *Quality Management in Urban Tourism*, Chichester: Wiley.

Hall, C. (2006) 'Urban entrepreneurship, corporate interests and sports mega-events: the thin policies of competitiveness within the hard outcomes of neoliberalism', *Sociological Review*, 54(2): 59–70.

Hall, P. (1998) *Cities in Civilization*, New York: Fromm International.

Hall, P. (2000) 'Creative cities and economic development', *Urban Studies*, 37(4): 639–49.

Hall, P. and Hall, P. (2006) 'Reurbanising the suburbs? The role of theatre, the arts and urban studies', *City*, 10(3): 377–92.

Hallam, E. and Ingold, T. (2007) 'Creativity and cultural improvisation: an introduction', in E. Hallam and T. Ingold (eds), *Creativity and Cultural Improvisation*, London: Routledge.

Halstead, R., Hazeley, J., Morris, A. and Morris, J. (2006) *Bollocks to Alton Towers: Uncommonly British Days Out*, London: Penguin.

Hannigan, J. (1999) *Fantasy City: Pleasure and Profit in the Postmodern Metropolis*, New York: Routledge.

Haraway, D. (2003) *The Companion Species Manifesto: Dogs, People and Significant Otherness*, Chicago: Prickly Paradigm Press.

Harrison, P. (2000) 'Making sense: embodiment and the sensibilities of the everyday', *Environment and Planning D: Society and Space*, 18: 497–517.

Harvey, D. (1989a) 'From managerialism to entrepreneurialism: the transformation in urban governance in late capitalism', *Geografiska Annaler*, 71(1): 3–17.

Harvey, D. (1989b) *The Condition of Post Modernity*, Oxford: Blackwell.

Harvey, D. (2008) 'The right to the city', *New Left Review*, 53: 23–40.

Hastings, A., Flint, J., McKenzie, C. and Mills, C. (2005) *Cleaning Up Neighbourhoods: Environmental Problems and Service Provision in Deprived Areas*, Bristol: Policy Press.

Haylett, C. (2003) 'Care, class and welfare reform: reading meanings, talking feelings', *Environment and Planning A*, 35(5): 799–814.

Hebdige, D. (1983) *Subculture: The Meaning of Style*, London: Methuen.

Hertzberger, H. (1991) *Lessons for Students in Architecture*, Rotterdam: Uitgiverij.

Hesmondhalgh, D. (1996) 'Flexibility, post-Fordism and the music industries', *Media, Culture & Society*, 18(3): 469–88.

Hesmondhalgh, D. (2002) *The Cultural Industries*, London: Sage.

Hewison, R. (1995) *Culture and Consensus: England, Art and Politics since 1940*, London: Methuen.

Hewitt, A. and Jordan, M. (2003) *I Fail to Agree*, Sheffield: Site Gallery.

Hill Strategies Research Inc. (2004) *A Statistical Profile of Artists in Canada: Based on the 2001 Census*, Hamilton, Ontario: Department of Canadian Heritage: 1–27.

Hill Strategies Research Inc. (2006) *Artists in Large Canadian Cities*, Ottawa: Council for the Arts.

Hirsch, P. (1972) 'Processing fads and fashions: an organization-set analysis of cultural industry systems', *The American Journal of Sociology*, 77(4): 639–59.

Hoggart, R. (1957) *The Uses of Literacy: Aspects of Working-Class Life, with Special Reference to Publications and Entertainments*, Harmondsworth, Middlesex: Penguin.

Hoggett, P. and Bishop, J. (1986) *Organizing Around Enthusiasms*, London: Comedia.

Holland, L. (2004) 'Diversity and connections in community gardens: a contribution to local sustainability', *Local Environment*, 9(3): 285–305.

Houpt, S. (2008) 'Artists' home finds unlikely saviour', *The Globe & Mail*, 24 March, R1–R2.

Howard, R. (2005) 'Toward a theory of the world wide web vernacular: the case for pet cloning', *Journal of Folklore Research*, 3(42): 323–60.

Hubbard, P. (2006) *City*, London: Routledge.

Hutchison, R. and Forrester, S. (1987) *Arts Centres in the United Kingdom*, London: Policy Studies Institute.

Hutton, T. (2006) 'Spatiality, built form, and creative industry development in the inner city', *Environment and Planning A*, 38(10): 1819–41.

Hutton, T. (2008) *The New Economy of the Inner City*, London: Routledge.

Hyde, L. (1979) *The Gift: Imagination and the Erotic Life of Property*, New York: Random House.

Ingold, T. (2007) *Lines: A Brief History*, London: Routledge.

IDM (Institute of Design Montréal) (2001) *Strategic Action Plan: Institute of Design Montréal Horizon*, Montréal: Le Groupe Stragesult.

Ironmonger, D. (1996) 'Counting outputs, capital inputs and caring labor: estimating gross household output', *Feminist Economics*, 2: 37–64.

Isaacs, P. (2008) 'Parkdale vs Queen West', *Eye Weekly*. Available at www.eyeweekly.com/blog/post/31513.

Italian Gnome Liberationist Group M.A.L.A.G. (Movimento Autonomo per la Liberazione delle Anime da Giardino). Available at www.malag.it/.

Jackson, M., Kabwasa-Green, F. and Herranz, J. (2006) *Cultural Vitality in Communities: Interpretation and Indicators*, Washington, DC: The Urban Institute.

Jackson, P. (1989) *Maps of Meaning*, London: Routledge.

Jacobs, J. (1961) *The Death and Life of Great American Cities*, New York: Random House.

Jacobs, J. (1969) *The Economy of Cities*, New York: Random House.

Jameson, F. (1984) 'Postmodernism, or the cultural logic of late capitalism', *New Left Review*, 146: 53–92.

Jenkins, R. (1975) 'Technology and the market: George Eastman and the origins of mass amateur photography', *Technology and Culture*, 1(16): 1–19.

Jessop, B. (1990) *State Theory: Putting the Capitalist State in its Place*, Cambridge: Polity Press.

Jessop, B. (1996) 'Interpretive sociology and the dialectic of structure and agency: reflections on Holmwood and Stewart's explanation and social theory', *Theory, Culture and Society*, 13(1): 119–28.

Jessop, B. (2001) 'Institutional (re)turns and the strategic-relational approach', *Environment and Planning A*, 33(7): 1213–37.

Jessop, B. (2004) 'Cultural political economy, the knowledge-based economy, and the state', in A. Barry and D. Slater (eds), *The Technological Economy*, London: Routledge.

Jones, M. (1997) 'Spatial selectivity of the state? The regulationist enigma and local struggles over economic governance', *Environment and Planning A*, 29(5): 831–64.

Jones, M. (1999) *New Institutional Spaces*, London: Jessica Kingsley.

Jones, S. (2002) 'Music that moves: popular music, distribution and network technologies', *Cultural Studies*, 16(2): 213–32.

Jonsson, T. (2006) 'Space invaders: there goes the neighbourhood', *In/Site*, 15: 37–9.

Jordan, T. (2002) *Activism! Direct Action, Hacktivism and the Future of Society*, London: Reaktion.

Jordison, S. and Kiernan, D. (2006) *The Idler Book of Crap Towns: The 50 Worst Places to Live in the UK*, London: Macmillan.

Jorgensen, A. and Keenan, R. (2008) Urban Wildscapes. Available at http://environment-room.co.uk/urbanwildscapes/thebook.html.

Jowell, T. (2005) 'Why should government support the arts?' *Engage*, 17: 5–8.

Katz, C. (2004) *Growing Up Global: Economic Restructuring and Children's Everyday Lives*, Minneapolis: University of Minnesota Press.

Kelly, O. (1984) *Community, Art and the State: Storming the Citadels*, London: Comedia.

Kent, K. (1976) 'Pueblo and Navajo weaving traditions and the Western world', in N. Grayburn (ed.), *Ethnic and Tourist Arts: Cultural Expressions from the Fourth World*, Berkeley: University of California Press.

Kester, G. (2004) *Conversation Pieces: Community and Communication in Modern Art*, Berkeley: University of California Press.

Kingwell, M. (2008) 'Toronto: justice denied', *The Walrus*, January/February: 1–5.

Kipfer, S. and Keil, R. (2002) 'Toronto Inc? planning the competitive city in the new Toronto', *Antipode*, 34(2): 227–64.

Kitchen, L., Marsden, T. and Milbourne, P. (2006) 'Nature, state and community: social forestry in the post-industrial British countryside', *Geoforum*, 37(5): 831–43.

Kohn, Margaret (2003) *Radical Space: Building the House of the People*, Ithaca, NY: Cornell University Press.

Komter, A. (ed.) (1996) *The Gift: An Interdisciplinary Perspective*, Amsterdam: Amsterdam University Press.

Komter, A. (2005) *Social Solidarity and the Gift*, Cambridge: Cambridge University Press.

Kong, L. (2000) 'Culture, economy and policy: trends and developments', *Geoforum*, 31: 385–90.

Kong, L., Gibson, C., Khoo, L.-M. and Semple, A.-L. (2006) 'Knowledges of the creative economy: towards a relational geography of diffusion and adaptation in Asia', *Asia Pacific Viewpoint*, 47(2): 173–94.

Kopytoff, I. (1986) 'The cultural biography of things: commoditization as process', in A. Appadurai (ed.), *The Social Life of Things: Commodities in Cultural Perspective*, Cambridge: Cambridge University Press.

Krätke, S. (2002) *Medienstadt: Urbane Cluster und globale Zentren der Kulturproduktion*, Opladen: Leske and Budrich.

Kriznik, B. (2004) *Forms of Local Resistance: No Al 22%*, Barcelona: Institute of Advanced Architecture of Catalonia. Available at www2.arnes.si/~uljfarh5/kriznik_noal22@.pdf.

Kwon, M. (2002) *One Place after Another: Site Specific Art and Locational Identity*, Boston, MA: MIT Press.

La Politique Culturelle du Québec: Notre Culture, Notre Avenir (1992) Québec: Ministère des Affaires Culturelles, Gouvernment du Québec.

Lacher, K.T., Meline, K.P., Birch, J., Brantley, M. and Carnathan, T. (1995) *The Spirit of Christmas: an Ethnographic Investigation of the Christmas Ritual*, Wichita, KS: Southwestern Marketing Association. Available at www.sbaer.uca.edu/research/swma/1995/pdf/11.pdf.

Laclau, E. (2005) *The Populist Reason*, London: Verso.

Lacy, S. (1995) *Mapping the Terrain: New Genre Public Art*, Seattle: Bay Press.

Landry, C. (2000) *The Creative City. A Toolkit for Urban Innovators*, London: Earthscan.

Landry, C. and Bianchini, F. (1995) *The Creative City*, Bournes Green: Comedia.

Larsen, L. and Fowle, K. (2005) 'Lunch hour: art, community, administered space, and unproductive activity', in T. Purves (ed.), *What We Want is Free: Generosity and Exchange in Recent Art*, New York: SUNY Press.

Lash, S. and Urry, J. (1994) *The Economies of Signs and Space*, London: Sage.

Latour, B. (1999) 'When things strike back: a possible contribution of "science studies" to the social sciences', *British Journal of Sociology*, 51(1): 107–23.

Latour, B. (2004) 'Why has critique run out of steam? From matters of fact to matters of concern', *Critical Inquiry*, 30(2): 225–48.

Lavie, S., Narayan, K. and Rosaldo, R. (eds) (1993) *Creativity/Anthropology*, London: Cornell University Press.

Lawler, S. (2005) 'Disgusted subjects: the making of middle class identities', in *Sociological Review*, 53(3): 429–46.

Lawson, J. (2005) *City Bountiful: A Century of Community Gardening in America*, Berkeley: University of California Press.

Lawson-Smith, M. (2008) *Reciting the City*. Available at www.plymouthartscentre.org/art/recitingthecity.html.

LDA (London Development Agency) (2006) *Strategies for Creative Spaces: London Case Study*, London: London Development Agency.

Leadbeater, C. and Miller, P. (2004) *The Pro-Am Revolution: How Enthusiasts Are Changing Our Economy and Society*, London: Demos.

Leadbeater, C. and Oakley, K. (1999) *The Independents: Britain's New Cultural Entrepreneurs*, London: Demos.

Lee, M. (1997) 'Relocating location: cultural geography, the specificity of place and the City of Habitus', in J. McGuigan (ed.), *Cultural Methodologies*, London: Sage.

Lee, R. (2000) 'Shelter from the storm? Geographies of regard in worlds of horticultural consumption and production', *Geoforum*, 31(2): 137–57.

Lee, T. (2004) 'Creative shifts and directions', *International Journal of Cultural Policy*, 10(3): 281–99.

Lefebvre, H. (1971 [1991]) *The Production of Space*, trans. Donald Nicholson-Smith, Oxford: Blackwell.

Lefebvre, H. (1991) *Critique of Everyday Life*, trans. J. Moore, London: Verso.

Lefebvre, H. (1996) 'The right to the city', trans. E. Kofman and E. Lebas, in *Lefebvre's Writings on Cities*, Oxford: Blackwell.

Lefebvre, H. (2004) *Rhythmanalysis: Space, Time and Everyday Life*, trans. Stuart Elden and Gerald Moore, London: Continuum.

Le Gates, R. and Stout, F. (eds) (2003) *The City Reader*, 3rd edn, London: Routledge.

Lerman, L. (2002) 'Art and community: feeding the artist, feeding the art', in D. Adams and A. Goldbard (eds), *Community Culture and Globalization*, New York: The Rockefeller Foundation.

Leslie, D. and Rantisi, N. (2006) 'Governing the design economy in Montréal, Canada', *Urban Affairs Review*, 40(5): 1–29.

Leslie, E. (2001) 'Tate Modern: a year of sweet success', *Radical Philosophy*, 109: 2–5.

Lewandowska, M. and Cummings, N. (2004) *Enthusiasts from Amateur Film Clubs*, Warsaw: Centrum Sztuki Współczesnej Zamek Ujadowski [Polish and English texts accompanying gallery exhibition].

Ley, D. (2003) 'Artists, aestheticisation and the field of gentrification', *Urban Studies*, 40(12): 2527–44.

Leyshon, A. (2003) 'Scary monsters? Software formats, peer-to-peer networks, and the spectre of the gift', *Environment and Planning D*, 21: 533–58.

Leyshon, A., Lee, R. and Williams, C. (eds) (2003) *Alternative Economic Spaces*, London: Sage.

Leyshon, A., Webb, P., French, S., Thrift, N. and Crewe, L. (2005) 'On the reproduction of the musical economy after the internet', *Media, Culture and Society*, 27(2): 177–209.

Lippard, L. (1997) *The Lure of the Local*, New York: The New Press.

Liu, A. (2004) *The Laws of Cool: Knowledge Work and the Culture of Information*, Chicago: Chicago University Press.

Lloyd, R. (2004) 'The neighbourhood in cultural production: material and symbolic resources in the new bohemia', *City and Community*, 3(4): 343–71.

Lloyd, R. (2006) *Neo-Bohemia: Art and Commerce in the Postindustrial City*, London: Routledge.

Londos, E. (2006) 'Kitsch is dead – long live garden gnomes', *Home Cultures*, 3: 293–306.

Longhurst, R. (2006) 'Plots, plants and paradoxes: contemporary domestic gardens in Aotearoa, New Zealand', *Social and Cultural Geography*, 7(4), 581–93.

Loukaitou-Sideris, A. and Banerjee, T. (1998) *Urban Design Downtown: Poetics and Politics of Form*, Berkeley: University of California Press.

Lucas, K., Fuller, S., Psaila, A. and Thrush, D. (2004) *Prioritising Local Environmental Concerns: Where There's A Will There's A Way*, York: Joseph Rowntree Foundation.

Lupton, R. and Power, A. (2002) 'Social exclusion and neighbourhoods' in J. Hills, J. Le Grand and D. Piachaud (eds), *Understanding Social Exclusion*, Oxford: Oxford University Press.

Luton, S. (2008) 'Is the public arts centre a waste of public money?', *Building Design*, London, 13 June.

Macalister, T. (2008) 'Invest in Iraq and you repeat past mistakes, investors tell BP board', *The Guardian*, 18 April, 29.

McBride, J. and Wilcox, A. (eds) (2006) *uTOpia: Towards A New Toronto*, Toronto: Coach House Books.

McCarthy, A. (2006) 'From the ordinary to the concrete: cultural studies and the politics of scale', in J. Schwoch and M. White (eds), *Questions of Method in Cultural Studies*, Malden, MA: Blackwell.

McGuigan, J. (1992) *Cultural Populism*, London: Routledge.

Machan, T.R. (1990) 'Politics and generosity', *Journal of Applied Philosophy*, 7(1): 61–73.

MacKenzie, A. (2005) 'The performativity of code: software and cultures of circulation', *Theory, Culture & Society*, 22(1): 71–92.

McLean, H. (2006) 'Go West young hipster: the gentrification of Queen Street West', in J. McBride and A. Wilcox, (eds), *uTOpia: Towards A New Toronto*, Toronto: Coach House Books.

McNeill, D. (2008) 'The hotel and the city', *Progress in Human Geography*, 32(3): 383–98.

McRobbie, A. (1998) *British Fashion Design: Rag Trade or Image Industry?*, London: Routledge.

McRobbie, A. (2002) 'Clubs to companies: notes on the decline of political culture in speeded up creative worlds', *Cultural Studies*, 16(4): 516–31.

McRobbie, A. (2005) '"Everyone is creative"; artists as pioneers of the new economy?' in J. Hartley (ed.), *Creative Industries*, Malden, MA: Blackwell.

Madanipour, A. (2003) *Public and Private Spaces of the City*, London: Routledge.

Malmberg, A. and Power, D. (2005) 'On the role of global demand in local innovation processes', in P. Shapiro and G. Fuchs (eds), *Rethinking Regional Innovation and Change*, New York: Springer.

Manske, A. (2006) 'Die Stellung halten. Marktstrategien und Positionskämpfe in Berlins Internetbranche', *Soziale Welt*, 57: 157–75.

Marcuse, P. (2002) 'The layered city', in P. Madsen and R. Plunz (eds), *The Urban Lifeworld: Formation, Perception, Representation*, New York: Routledge.

Markus, T. (1994) *Buildings and Power: Freedom and Control in the Origin of Modern Building Types*, London: Routledge.

Markusen, A. (1996) 'Sticky places in slippery space: a typology of industrial districts', *Economic Geography*, 72(3): 293–313.

Markusen, A. (2004) 'An actor-centered approach to economic geographic change', *Annals of the Japan Association of Economic Geographers*, 49(5): 395–408.

Markusen, A. (2006) 'Building the creative economy for Minnesota's artists and communities', *Cura Reporter*, Summer: 16–25.

Markusen, A. and Johnson, A. (2006) *Artists' Centers: Evolution and Impact on Careers, Neighborhoods, and Economies*. Minneapolis: Project on Regional and Industrial Economics, University of Minnesota.

Markusen, A. and King, D. (2003) 'The artistic dividend: the hidden contributions of the arts to the regional economy. Project on Regional and Industrial Economics', Minneapolis: Humphrey Institute, University of Minnesota.

Markusen, A., Rendon, M. and Martinez, A. (2008) 'Native American artists and their gatekeepers and markets: a reflection on regional trajectories', Working paper no. 275, Project on Regional and Industrial Economics, Minneapolis: Humphrey Institute of Public Affairs, University of Minnesota.

Markusen, A., Wassall, G., DeNatale, G. and Cohen, R. (2008) 'Defining the creative economy: industry and occupational approaches', *Economic Development Quarterly*, 22(1): 24–45.

Markusen, A., Gilmore, S., Johnson, A., Levi, T. and Martinez, A. (2006) *Crossover: How Artists Build Careers across Commercial, Nonprofit and Community Work*, Minneapolis: Project on Regional and Industrial Economics, University of Minnesota.

Maskell, P. and Malmberg, A. (1999) 'Localised learning and industrial competitiveness', *Cambridge Journal of Economics* 23(2): 167–85.

Mass Observation (1943) 'Public taste and public design', File Report 1675, 8 May.

Massey, D. (1994) *Space, Place and Gender*, Cambridge: Polity Press.

Massey, D. (1998) 'The spatial construction of youth cultures', in T. Skelton and G. Valentine (eds), *Cool Places: Geographies of Youth Cultures*, London: Routledge.

Massey, D. (2005) *For Space*, London: Sage.

Matarasso, F. (1999) *Towards a Local Culture Index: Measuring the Cultural Vitality of Communities*, Stroud: Comedia.

Mattin (2005) 'Give it all, zero for rules!', *Mute Magazine*, 29 March. Available at www. metamute.org/en/Give-It-All-Zero-For-Rules.

Mauss, M. (1967) *The Gift: Forms and Functions of Exchange in Archaic Societies*, trans. Ian Cunnison, New York: Norton.

Meinig, D. (1979) (ed.) *The Interpretation of Ordinary Landscapes*, New York: Oxford University Press.

Mele, C. (2000) *Selling the Lower East Side: Culture, Real Estate, and Resistance in New York, 1880–2000*, Minneapolis: University of Minnesota Press.

Melucci, A. (1989) *Nomads of the Present: Social Movements and Individual Needs in Contemporary Society*, London: Hutchinson Radius.

Menger, P.-M. (2006) *Kunst und Brot: Die Metamorphosen des Arbeitnehmers*, Konstanz: UVK Verlagsgesellschaft.

Meredith, S. (2005) 'Fatal attraction for divers', *News & Star*, 12 February. Available at www.gnomeland.co.uk/News-LAKES%20GNOME%20GARDEN.html.

Michalski, S. (1998) *Public Monuments: Art in Political Bondage 1870–1997*, London: Reaktion.

Miège, B. (1989) *The Capitalization of Cultural Production*, New York: International General.

Miles, M. (2004) *Urban Avant-Gardes: Art, Architecture and Change*, London: Routledge.

Miles, M. (2005) 'Interruptions: testing the rhetoric of culturally-led urban development', *Urban Studies*, 42(5/6): 889–911.

Miles, M. (2007) *Cities and Cultures*, London: Routledge.

Miles, M., Hall, T. and Borden, I. (eds) (2003) *The City Cultures Reader*, London: Routledge.

Miles, S. (2005) 'Creativity, culture and urban development: Toronto examined', *disP* 163(3): 70–87.

Miles, S. and Paddison, R. (2005) 'Introduction: the rise and rise of culture-led urban regeneration', *Urban Studies*, 42: 833–39.

Miller, D. (ed.) (1993) *Unwrapping Christmas*, Oxford: Oxford University Press.

Milroy, S. (2008) 'The promised land?', *The Globe & Mail*, 16 January, R4.

Mindell, A. (1992) *The Leader as Martial Artist: An Introduction to Deep Democracy*, San Francisco: Harper.

Minus, J. (2007) 'An economy all shook up', *The Australian*, 6 January, 6.

Mitchell, D. (2000) *Cultural Geography: A Critical Introduction*, Blackwell: Oxford.

Molotch, H. (1996) 'L.A. as design product', in A.J. Scott and E.W. Soja (eds), *The City: Los Angeles and Urban Theory at the End of the Twentieth Century*, Berkeley: University of California Press, 225–75.

Molotch, H. (2003) *Where Stuff Comes From: How Toasters, Toilets, Cars, Computers, and many other things come to be as they are*, New York: Routledge.

Mommaas, H. (2004) 'Cultural cluster and the post-industrial city: towards the remapping of urban cultural policy', *Urban Studies*, 41: 507–32.

Montgomery, J. (2005) 'Beware the creative class: creativity and wealth creation revisited', *Local Economy*, 20(4): 337–43.

Montgomery, J. (2008) *The Newest Wealth of Cities: City Dynamics and the Fifth Wave*, Aldershot: Ashgate.

Moore, L. (2001) 'Elle meets the president: weaving Navajo culture and commerce in the Southwestern tourism industry', *Frontiers: A Journal of Women Studies*, 22(1): 21–44.

Morris, J. and Urry, J. (2006) *Growing Places: A Study of Social Change in The National Forest*, Farnham: Forest Research.

Mowl, T. (2007) *Historic Gardens of Oxfordshire*, Stroud: Tempus Publishing.

Mundelius, M. (2006) *Die Bedeutung der Kulturwirtschaft für den Wirtschaftsstandort Pankow*, Berlin: DIW Berlin.

Muñoz, F. (2006) 'Olympic urbanism and Olympic villages: planning strategies in Olympic host cities, London 1908 to London 2012', *Sociological Review*, 53: 175–87.

Muñoz, F. (1998) *Generosity: Virtue in Civil Society*, Washington, DC: Cato Institute.

Muttitt, G. and Marriott, J. (2002) *Some Common Concerns: Imagining BP's Azerbaijan-Georgia-Turkey Pipelines System*, London: published jointly by Platform, The Corner house, Friends of the Earth International, Campagna per la Riforma della Banca Mondiale, CEE Bankwatch, and Kurdish Human Rights Project.

Nachmanovitch, S. (1990) *Free Play: Improvisation in Life and Art*, New York: G.P. Putnam's Sons Ltd.

Nanni, L. and Houston, A. (2006) 'Editorial: Heterotopian creation: beyond the Utopia of theatres and galleries – site specific performance', *Canadian Theatre Review*, 126: 5–9.

National Economics (2002) *The State of the Regions*, Melbourne: National Economics and the Australian Local Government Association.

Neergaard, L. (2008) 'Capturing creativity?', *The Globe & Mail*, 13 March, A2.

Neff, G., Wissinger, E. and Zukin, S. (2005) 'Entrepreneurial labor among cultural producers: "cool" jobs in "hot" industries', *Social Semiotics*, 15(3): 307–34.

Negus, K. (1999) *Music Genres and Corporate Cultures*, London: Routledge.

Nek Chand Foundation (2009) Available at www.nekchand.com/index.html.

Nicolaides, B. (2006) 'How hell moved from the city to the suburbs: urban scholars and changing perceptions of authentic community', in K. Kruse and T. Sugrue (eds), *The New Suburban History*, Chicago: The University of Chicago Press.

Noble, M. (2008) *Lovely Spaces in Unknown Places: Creative City Building in Toronto's Inner Suburbs*, Toronto: Cities Centre, University of Toronto.

Oakes, L. (2006) 'Ojibwe scrolls come full circle', *Minneapolis Star-Tribune*, December.

Oakley, K. (2006) 'Include us out – economic development and social policy in the creative industries', *Cultural Trends*, 15(4): 255–73.

O'Connor, J. (1999) 'The definition of "cultural industries"', Manchester: Manchester Institute for Popular Culture. Available at http://mmu.ac.uk/h-ss/mipc/iciss/home2.htm.

O'Connor, J. (2004) '"A special kind of city knowledge": innovative clusters, tacit knowledge and the "creative city"', *Media International Australia incorporating Culture and Policy*, 112: 131–49.

O'Donnell, D. (2006) *Social Acupuncture: A Guide to Suicide, Performance and Utopia*, Toronto: Coach House Books.

O'Reilly, T. (2005) 'What is Web 2.0? design patterns and business models for the next generation of software', *O'Reilly Network*. Available at www.oreillynet.com/pub/a/oreilly/tim/news/2005/09/30/what-is-web-20.html.

ODPM (Office of the Deputy Prime Minister) (2003) *Sustainable Communities Plan*, www.odpm.gov.uk/stellent/groups/odpm_communities/documents/page/odpm_comm_022208.hcsp.

ODPM (2004) *Competitive Cities make Prosperous Regions and Sustainable Communities*, Third Report of the Core Cities Working Group, London: HMSO.

Oliver, P., Davis, I. and Bentley, I. (1994) *Dunroamin: Suburban Semi and Its Enemies*, London: Pimlico.

Operation Centaur Rodeo (2007) http://hoofinthemouth.blogspot.com/2007_09_09archive.html.

Osbourne, T. (2003) 'Against "creativity": a philistine rant', *Economy and Society*, 32(4): 507–25.

Osteen, M. (2002) *The Question of the Gift*, London: Routledge.

Owens, A. (2006) 'Book review: Geographies of British modernity: space and society in the twentieth century', *Progress in Human Geography*, 30: 266–8.

Park, R., Burgess, E. and McKenzie, R. (1967) [1925] *The City*, Chicago: Chicago University Press.

Pearson, D. (2008) 'Magical mystery tour', *Observer*, 17 August. Available at www.guardian.co.uk/lifeandstyle/2008/aug/17/gardens.georgeharrison.

Pearson, M. and Shanks, M. (2001) *Theatre/Archaeology*, Routledge: London.

Peck, J. (2004) 'Geography and public policy: constructions of neoliberalism', *Progress in Human Geography*, 28(3): 392–405.

Peck, J. (2005) 'Struggling with the creative class', *International Journal of Urban and Regional Research*, 29(4): 740–70.

Perl, J. (2001) 'The adolescent city', *The New Republic*, 22 January, 23–30.

Perrone, C. and Dunn, C. (eds) (2001) *Brazilian Popular Music and Globalization*, Gainesville, FL: University Press of Florida.

Perroux, F. (1950) 'Economic spaces: theory and applications', *Quarterly Journal of Economics*, 64(1): 89–104.

Peterson, E. (1996) *The Changing Faces of Tradition: A Report on the Folk and Traditional Arts in the United States*. Washington, DC: National Endowment for the Arts, Research Division Report no. 38.

Picard, L. and de Cortret, R.R. (1986) *Report of the Consultative Committee to the Ministerial Committee on the Development of the Montréal Region*, Ottawa: Minister of Supply and Services Canada.

Pick, J. (1991) *Vile Jelly: The Birth, Life and Lingering Death of the Arts Council of GB*, Doncaster: Brymill.

Pile, S. (1999) 'The heterogeneity of cities', in S. Pile, C. Brook and G. Mooney (eds), *Unruly Cities?*, London: The Open University Press.

Pile, S. (2005) *Real Cities, Space and the Phantasmagorias of City Life*, London: Sage.

Pinder, D. (2005) 'Arts of urban exploration', *Cultural Geographies*, 12: 383–411.

Piore, M.J. and Sabel, C.F. (1984) *The Second Industrial Divide: Possibilities for Prosperity*, Basic Books: New York.

Polese, M. and Shearmur, R. (2004) 'Culture, language, and the location of high-order service functions: the case of Montréal and Toronto', *Economic Geography*, 80(4): 329–50.

Pollay, R.W. (1987) 'It's the thought that counts: a case study in Xmas excesses', *Advances in Consumer Research*, 14(1): 140–43.

Porter, M. (1985) *Competitive Advantage*, The Free Press: New York.

Porter, M. (1998) 'Clusters and the new economics of competition', *Harvard Business Review*, November–December, 77–90.

Potts, T. (2007) '"Walking the line": kitsch, class and the morphing subject of value', Nottingham: Nottingham Modern Languages Press. Available at http://mlpa.nottingham.ac.uk/archive/00000071/.

Power, D. and Hallencreutz, D. (2005) 'Competitiveness, local productions systems and global commodity chains in the music industry: entering the US market', *Centre for Research on Innovation and Industrial Dynamics*, Uppsala: Uppsala Universitet: 1–22.

Pratt, A. (2004) 'Creative clusters: towards the governance of the creative industries production system', *Media International Australia*, 112: 50–66.

Pratt, A. (2005) 'Cultural industries and public policy: an oxymoron?' *International Journal of Cultural Policy*, 11(1): 31–44.

Pratt, A. (2008) 'Cultural commodity chains, cultural clusters, or cultural production chains?', *Growth and Change*, 39(1): 95–103.

Projekt Zukunft (2005) *Kulturwirtschaft in Berlin: Entwicklung und Potenziale*, Berlin: Geschäftsstelle Projekt Zukunft.

Projekt Zukunft (2006) *Flyer Creative Industries Initiative: Creative Industries Berlin*, Berlin: Geschäftsstelle Projekt Zukunft.

Purcell, M. (2002) 'The state, regulation, and global restructuring: reasserting the political in political economy', *Review of International Political Economy*, 9(2): 298–332.

Purves, T. (2005) *What We Want is Free: Generosity and Exchange in Recent Art*, New York: SUNY Press.

Radley, A. (1995) 'The elusory body and social constructionist theory', *Body and Society*, 1(2): 3–23.

Raffel, S. (2001) 'On generosity', *History of Human Sciences*, 14(8): 111–28.

Raffo, C., O'Connor, J., Lovatt, A. and Banks, M. (2000) 'Risk and trust in the cultural industries', *Geoforum*, 31(4): 453–64.

Reuters (2000) 'Garden Gnome Liberation Front strikes Paris show'. Available at http://archives.cnn.com/2000/STYLE/arts/04/12/france.gnomes.reut/.

Reynolds, S. (1998) *Energy Flash: A Journey through Rave Music and Dance Culture*, London and Basingstoke: Picador.

RHS (Royal Horticultural Society) (2007) 'General show regulations for exhibitors', London: Shows Department, Royal Horticultural Society, 1–48.

Richardson, G. (1972) 'The organisation of industry', *The Economic Journal*, 82(327): 883–96.

Robbins, E. (1996) 'Thinking space/seeing space: Thamesmead revisted', *Urban Design International*, 1(3): 283–91.

Román, A. (2006) 'Bilbao: a spectacular but somehow disenchanted city', paper presented at Drawing the Lines: International Perspectives on Urban Renewal Through the Arts, Indiana University Northwest, Gary, Indiana, 2–4 November.

Rose, C. (1989) 'The comedy of the commons: commerce, custom and inherently public property', *University of Chicago Law Review*, 53: 711–18.

Routledge, P. (1997) 'Pollock Free State and the practice of postmodern politics', *Transactions of the Institute of British Geographers*, 22: 359–77.

Russell, V. (2005) *Gnomes*, London: Frances Lincoln.

Ruting, B. and Li, J. (2008) *Parkes Elvis Festival 2008: Preliminary Report*, Sydney: University of Sydney.

Ryan, B. (1992) *Making Capital from Culture: The Corporate Form of Capitalist Cultural Production*, Berlin: Walter de Gruyter.

Sack, R. (1997) *Homo Geographicus*, Baltimore: Johns Hopkins University Press.

Salford City Council (2009) Available at www.salford.gov.uk/housing/marketrenewal/seedlang-srb/seedleylang-srb-history-and-background.htm.

Salford Star (2006) 'Lifting the lid on the Lowry', 12 August. Available at http://salford-star.blogspot.com2006/08/lifting-lid-on-lowry.

Sandercock, L. (1998) *Towards Cosmopolis*, Chichester: Wiley.

Sandercock, L. (2006) 'Cosmopolitan urbanism: a love song to our mongrel cities', in J. Binnie, J. Holloway, S. Millington, and C. Young (eds), *Cosmopolitan Urbanism*, London: Routledge.

Santagata, W. (2004) 'Creativity, fashion and market behavior', in D. Power, and A. Scott (eds), *Cultural Industries and the Production of Culture*, London: Routledge.

Sarmiento, C. (2006) 'A study of radical and transnational space: el Centro Cultural de Mexico', paper presented at the Annual American Collegiate Schools of Planning Meetings, Forth Worth, Texas, October.

Sassen, S. (2006) *Cities in a World Economy*, Thousand Oaks, CA: Pine Forge Press.

Savage, M. and Warde, A. (1993) *Urban Sociology: Capitalism and Modernity*, Basingstoke: Macmillan.

Schmelzkopf, K. (1995) 'Urban community gardens as contested space', *Geographical Review*, 85(3): 364–81.

Schouvaloff, A. (ed.) (1970) *Place for the Arts*, Liverpool: North West Arts Association/Seel House.

Schrift, A. (ed.) (1997) *The Logic of the Gift: Toward an Ethic of Generosity*, London: Routledge.

Scott, A. (1996) 'The craft, fashion, and cultural products industries of Los Angeles: competitive dynamics and policy dilemmas in a multi-sectoral image-producing complex', *Annals of the Association of American Geographers*, 86(2): 306–23.

Scott, A. (1997) 'The cultural economy of cities', *International Journal of Urban and Regional Research*, 2(1): 323–39.

Scott, A. (1999) 'The cultural economy: geography and the creative field', *Media, Culture and Society*, 21: 807–17.

Scott, A. (2000) *The Cultural Economy of Cities*, London: Sage.

Scott, A. (2006) 'Creative cities: conceptual issues and policy questions', *Journal of Urban Affairs*, 28(1): 1–18.

Scott, A. (2008) *Social Economy of the Metropolis: Cognitive-Cultural Capitalism and the Global Resurgence of Cities*, Oxford: Oxford University Press.

Scuba Gnomes. Available at www.gordonmackie.com/scuba_gnomes.htm#picture.

Seddon, G. (1997) *Landprints: Reflections on Place and Landscape*, Cambridge: Cambridge University Press.

Seedley and Langworthy Trust (2009) www.seedleytrust.co.uk.

Shaw, P. (1999) *The Arts and Neighbourhood Renewal: A Research Report*, Policy Action Team 10, London: Department for Culture, Media and Sport.

Shaw, S. (2007) 'Hosting a sustainable visitor economy: lessons from the regeneration of London's "Banglatown"', *Journal of Urban Regeneration and Renewal*, 1(3): 275–85.

Sheppard, E. (2002) 'The spaces and times of globalization: place, scale, networks, and positionality', *Economic Geography*, 78(3): 307–30.

Short, J. (1996) *The Urban Order: An Introduction to Cities, Cultures, and Power*, Oxford: Blackwell.

Siegle, L. (2005) 'The A–Z of gardening', *The Observer*, 26 June. Available at www.guardian.co.uk/lifeandstyle/2005/jun/26/shopping.ethicalliving.

Silk, J. (1998) 'Caring at a distance', *Ethics, Place and Environment*, 3(3): 303–22.

Silverstone, R. (1997) 'Introduction', in R. Silverstone (ed.) *Visions of Suburbia*, London: Routledge.

Simpkins, D. (2008) 'Future visions of history'. Available at www.danielsimpkins.net/main/future/future1.html.

Skeggs, B. (1997) *Formations of Class and Gender: Becoming Respectable*, London: Sage.

Skeggs, B. (2004) *Class, Self, Culture*, London: Routledge.

Slater, D. (1991) 'Consuming Kodak', in J. Spence and P. Holland (eds), *Family Snaps: The Meaning of Domestic Photography*, London: Virago.

Slater, T. (2004) 'Municipally managed gentrification in South Parkdale, Toronto', *The Canadian Geographer*, 48(3): 303–25.

Smith, N. (1996) *The New Urban Frontier: Gentrification and the Revanchist City*, Routledge: London.

Smith, N. and Low, S. (eds) (2006) *The Politics of Public Space*, London: Routledge.

Soja, E. (1996) *Thirdspace: Journeys to Los Angeles and Other Real and Imagined Places*, Oxford: Blackwell.

Somers, M. (2005) 'Sociology and economics. Beware trojan horses bearing social capital: how privatization turned "solidarity" into a bowling team', in G. Steinmetz (ed.), *The Politics of Method in the Human Sciences: Positivism and its Methodological Others*, London: Duke University Press.

St Martin's Lane Hotel. Available at: www.stmartinslane.com/#/home/.

Stadtforum Berlin (2006) *Perspektiven für Berlin: Strategien und Leitprojekte*, Berlin: Senatsverwaltung für Stadtentwicklung, 164–77.

Stanes, E., Mansfield, J., Gibson, C., Brennan-Horley, C. and Stewart, A. (2007) *Parkes Elvis Revival Festival: Visitor Survey Results 2007*, Parkes: Parkes City Council.

Stanwick, S. and Flores, J. (2007) *Design City: Toronto*, Chichester: John Wiley & Sons.

Sternad, J. (n.d.) Available at www.latinart.com/transcript.cfm?id=88.

Stocker, L. and Barnett, K. (1998) 'The significance and praxis of community-based sustainability projects', *Local Environment*, 3(2): 179–91.

Stolarick, K. and Florida, R. (2006) 'Creativity, connections and innovation: a study of linkages in the Montréal region', *Environment and Planning A*, 38(10): 1799–817.

Straw, W. (1991) 'Systems of articulation, logics of change: communities and scenes in popular music', *Cultural Studies*, 5(3): 368–88.

Straw, W. (2002) 'Value and velocity: the 12-inch single as medium and artifact', in D. Hesmondhalgh and K. Negus (eds), *Popular Music Studies*, London: Arnold.

Street Porter, J. (2001) 'LifeEtc', *Independent on Sunday*, 21 October, 2.

Sustainable Development Commission (2002) *Vision for Sustainable Regeneration, Environment and Poverty – The Missing Link*, London: Sustainable Development Commission.

Taylor, C. (1988) 'Various approaches to and definitions of creativity', in R. Sternberg (ed.), *The Nature of Creativity*, Cambridge: Cambridge University Press.

Taylor, D. (2003) *The Archive and the Repertoire. Performing Cultural Memory in the Americas*, Durham, NC: Duke University Press.

Taylor, J. (2008) 'Writing is on the wall for council graffiti plan', *Metro*, 16 October, 27.

Temporary Services (eds) (2007) *Group Work*, New York: Printed Matter, Inc.

Terranova, T. (2004) *Network Culture: Politics for the Information Age*, London: Pluto Press.

Thompson, N. and Sholette, G. (eds) (2004) *Interventionists: Users' Manual for the Creative Disruption of Everyday Life*, Cambridge, MA: MIT Press.

Thornton, S. (1996) *Club Cultures: Music, Media and Subcultural Capital*, Middletown, Wesleyan University Press.

Thorsby, D. (2001) *Economics and Culture*, Cambridge: Cambridge University Press.

Thrift, N. (2008) *Non-representational Theory*, London: Routledge.

Tisdale, S. (1996) 'Railroads, tourism, and Native Americans in the Greater Southwest', *Journal of the Southwest*, 38(4): 433–62.

Topping, A. (2008) 'Party across the Mersey', *The Guardian*, 9 January. Available at www. guardian.co.uk/society/2008/09/localgovernment.regeneration.

Townsend, P. (ed.) (1983) *Art Within Reach*, London: Thames and Hudson.

Travelocity Roaming Gnome. Available at http://leisure.travelocity.com/Promotions/0,,TR AVELOCITY|1751|mkt_main|,00.htm.

Turok, I. (2003) 'Cities, clusters and creative industries: the case of film and television in Scotland', *European Planning Studies*, 11(5): 549–65.

Tusa, J. (2004) *On Creativity: Interviews Exploring the Process*, London: Methuen.

Tyler, I. (2008) '"Chav mum, chav scum": class disgust in contemporary Britain', *Feminist Media Studies*, 8(1): 17–34.

Ursell, G. (2000) 'Television production: issues of exploitation, commodification and subjectivity in UK television labour markets', *Media, Culture & Society*, 22(6): 805–25.

Van Heur, B. (2009a) 'From creative industries to critique: comparing policies on London and Berlin', in F. Eckardt and L. Nyström (eds), *Culture and the City*, Berlin: Berliner Wissenschaftsverlag.

Van Heur, B. (2009b) 'The clustering of creative networks: between myth and reality', *Urban Studies*, 46(8): 1531–52.

Vaughan, G. (1997) *For-giving: A Feminist Criticism of Exchange*, Austin: Plain View Press.

Vaughan, G. (ed.) (2004) *Il Dono / The Gift: A Feminist Analysis*, London: Meltemi Press.

Vaughan, G. (ed.) (2007) *Women and the Gift Economy: A Radically Different Worldview is Possible*, Toronto: Inanna Publications and Education.

Veblen, T. (1899, 1994) *The Theory of the Leisure Class*, in *The Collected Works of Thorstein Veblen, Vol. 1*, London: Routledge.

Veloso, C. (2002) *Tropical Truth: The Story of Music and Revolution in Brazil*, New York: Alfred A. Knopf.

Vicario, L. and Martínez-Monje, P.M. (2006) 'The "Guggenheim Effect" and the "New Bilbao": on the social costs of Bilbao's urban regeneration', paper presented at Drawing the Lines: International Perspectives on Urban Renewal Through the Arts, Indiana University Northwest, Gary, Indiana, 2–4 November.

Wade, E. (1974) 'Change and development in the Southwest Indian art market', in L. Eiseley, J. Beal, E. Wade, D. Shwartz, and D. Noble, *Exploration: Annual Bulletin of the School of American Research*, 16–21.

Waitt, G. (2008) 'Urban festivals: geographies of hype, helplessness and hope', *Geography Compass*, 2(2): 513–37.

Wali, A., Severson, R. and Longoni, M. (2002) *Informal Arts: Finding Cohesion, Capacity and Other Cultural Benefits in Unexpected Places*, Chicago: Chicago Center for Arts Policy at Columbia College.

Walker, G. and Bickerstaff, K. (2000) 'Polluting the poor: an emerging environmental justice agenda for the UK', CUCR paper, London: Goldsmiths College, University of London.

Walker, O. (2008) Postcard publicising *Mr Democracy.* Available at www.mrdemocracy.org.

Walks, R. (2001) 'The social ecology of the post-Fordist/global city? Economic restructuring and socio-spatial polarisation in the Toronto urban region', *Urban Studies*, 38(3): 407–47.

Wall, D. (1999) *Earth First! And the Anti-Roads Movement*, London: Routledge.

Wang, D. (2004) 'MESS HALL: What it is (after the first year)'. Available at www.messhall.org/wimh.html.

Ward, D. (2004) 'How an avocado bathroom can slash your home's value by £8,000', *The Guardian*, 22 July. Available at www.guardian.co.uk/money/2004/jul/22/houseprices.homeimprovements.

Warpole, K. (2000) *In Our Backyard: The Social Promise of Environmentalism*, London: Groundwork and Green Alliance.

Webster, L. (1996) 'Reproducing the past: revival and revision in Navajo weaving', *Journal of the Southwest*, 38(4): 415–31.

Wellingborough Borough Council (2007) *Results from MORI Best Value Reports*, Community Committee, 3 December.

Wellman, B. (2001) 'Physical place and cyberspace: the rise of networked individualism', *International Journal of Urban and Regional Research*, 25: 227–52.

Werbner, P. (1999) 'Global pathways, working class cosmopolitans and the creation of trans ethnic worlds', *Social Anthropology*, 7(1): 17–35.

West, C. (2007) Weblog. Available at www.bbc.co.uk/gardening/today_in_your_garden/weblogs/weblog_clevewest/200709/19421.shtml.

Whyte, W. (1956) *The Organization Man*, New York: Doubleday.

Wiles, W. (2007) 'Kitsch is a controversial subject. "Sometimes it upsets people"', *Icon*, 51 (September).

Williams, R. (1958a) *Culture and Society 1780–1950*, London: Chatto & Windus.

Williams, R. ([1958b] 1997) 'Culture is ordinary', in A. Gray and J. McGuigan (eds) *Studies in Culture: An Introductory Reader*, London: Arnold.

Williams, R. (1961) *The Long Revolution*, London: Pelican.

Willis, P. (1991) *Towards a New Cultural Map*, consultation publication on the British National Arts and Media Strategy. London: Arts Council of Great Britain.

Willis, P. (1993) 'Symbolic creativity', reproduced in A. Gray. and J. McGuigan (eds), *Studies in Culture: An Introductory Reader*, London: Arnold, 208–16.

Wilson, E. (1991) *The Sphinx in the City*, London: Virago.

Wilson, E. (2003) *Bohemians: The Glamorous Outcasts*, London: Tauris Parke.

Winchester, H.P.M. and Rofe, M.W. (2005) 'Christmas in the "Valley of Praise": intersections of the rural idyll, heritage and community in Lobethal, South Australia', *Journal of Rural Studies*, 21: 265–79.

Wirth, L. (2003) 'Urbanism as a way of life', in R. LeGates and F. Stout (eds) *The City Reader*, London: Routledge, 156–63.

Wolfe, D. and Gertler, M. (2004) 'Clusters from the inside and out: local dynamics and global linkages', *Urban Studies*, 41(5/6): 1071–93.

Wyszomirski, M. (2000) 'Raison d'état, raisons des arts: thinking about public purposes', in J. Cherbo and M. Wyszomirski (eds), *The Public Life of the Arts in America*, New Brunswick: Rutgers University Press.

Young, I. (1990) *Justice and the Politics of Difference*, Princeton: Princeton University Press.

Yúdice, G. (2003) *The Expediency of Culture: Uses of Culture in the Global Era*, Durham, NC: Duke University Press.

Zeitlyn, D. (2003) 'Gift economies and the development of open source software: anthropological reflections', *Research Policy*, 32(7): 1287–91.

Zukin, S. (1982) *Loft Living: Culture and Capital in Urban Change*, New Brunswick: Rutgers University Press.

Zukin, S. (1991) *Landscapes of Power: From Detroit to Disney World*, Berkeley, CA: University of California Press.

Zukin, S. (1995) *The Cultures of Cities*, Oxford: Blackwell.

Zukin, S. (2006) 'An idea whose time has come? Authenticity and standardization in cultural strategies of redevelopment', paper presented at Drawing the Lines: International Perspectives on Urban Renewal Through the Arts, Indiana University Northwest, Gary, Indiana, 2–4 November.

Zukin, S. and Kosta, E. (2004) 'Bourdieu off-Broadway: managing distinction on a shopping block in the East Village', *City and Community*, 3(2): 101–14.

Index

Milton Keynes UK
Ingram Content Group UK Ltd.
UKHW031146141024
449569UK00024B/1025